G. Tomassini (Ed.)

Algebraic Surfaces

Lectures given at a Summer School of the
Centro Internazionale Matematico Estivo (C.I.M.E.),
held in Cortona (Arezzo), Italy,
June 22-30, 1977

C.I.M.E. Foundation
c/o Dipartimento di Matematica "U. Dini"
Viale margagni n. 67/a
50134 Firenze
Italy
cime@math.unifi.it

ISBN 978-3-642-11086-3 e-ISBN: 978-3-642-11087-0
DOI:10.1007/978-3-642-11087-0
Springer Heidelberg Dordrecht London New York

Reprint of the 1st ed. C.I.M.E., Ed. Liguori, Napoli 1981
With kind permission of C.I.M.E.

Printed on acid-free paper

Springer.com

C O N T E N T S

CENTRO INTERNAZIONALE MATEMATICO ESTIVO

(C.I.M.E.)

SURFACES ALGEBRIQUES COMPLEXES

ARNAUD BEAUVILLE

Surfaces Algébriques Complexes

Arnaud Beauville

INTRODUCTION

Cet exposé comprend deux parties. La première est un survol assez rapide de la classification d'Enriques des surfaces algébriques. On s'est inspiré, bien entendu, de la littérature classique sur le sujet, et en particulier du séminaire Chafarevitch [Ch.2]. On a essayé d'être aussi élémentaire que possible, en supposant toutefois connue la cohomologie des faisceaux cohérents. On renvoie à [Be] pour une exposition plus détaillée ainsi que pour des exemples.

La seconde partie comprend des indications sur la démonstration par Chafarevitch et Piatechki-Chapiro du théorème de Torelli pour les surfaces K 3 ([Ch.P]).

PREMIERE PARTIE

§1. Notations et rappels.

Nous dirons simplement surface au lieu de surface projective et lisse sur $\mathbb{C}$.

Soit S une surface, D, D' deux diviseurs sur S . On note :

- $D \equiv D'$ si D et D' sont linéairement équivalents (i.e. D-D' est le diviseur d'une fonction rationnelle sur S).
- $\mathcal{O}_S(D)$ le faisceau inversible associé à D .
- $H^i(S, \mathcal{O}_S(D))$, ou simplement $H^i(D)$ s'il n'y a pas de confusion possible, les espaces de cohomologie du faisceau $\mathcal{O}_S(D)$.
- $h^i(D) = \dim_{\mathbb{C}} H^i(D)$
- $\chi(\mathcal{O}_S(D)) = h^o(D) - h^1(D) + h^2(D)$
- $|D|$ = espace projectif des diviseurs effectifs linéairement équivalents à D .

  = espace projectif associé à $H^o(D)$.

  (Si $h^o(D) = 2$, on dit que $|D|$ est un pinceau).
- K_S ou K = diviseur canonique = un diviseur tel que $\mathcal{O}_S(K) = \Omega_S^2$.
- Pic(S) = groupe des diviseurs modulo équivalence linéaire

  $\cong$ groupe des classes d'isomorphisme de faisceaux inversibles

En vertu de théorèmes généraux, on a :

$$\mathrm{Pic}(S) = H^1(S, \mathcal{O}_S^*) = H^1(S, {}^h\mathcal{O}_S^*)$$

où ${}^h\mathcal{O}_S$ désigne le faisceau des fonctions holomorphes sur S, considérée comme variété analytique. Cette dernière interprétation permet de considérer la suite exacte :

$$0 \longrightarrow \mathbb{Z} \longrightarrow {}^h\mathcal{O}_S \xrightarrow{\exp} {}^h\mathcal{O}_S^* \longrightarrow 0$$

d'où l'on déduit la suite exacte importante :

$$0 \longrightarrow H^1(S,\mathbb{Z}) \longrightarrow H^1(S, \mathcal{O}_S) \longrightarrow \mathrm{Pic}(S) \longrightarrow H^2(S,\mathbb{Z}) \longrightarrow H^2(S, \mathcal{O}_S) \qquad (a)$$

Posons : $\mathrm{Pic}^\circ(S) = H^1(S, \mathcal{O}_S)/H^1(S,\mathbb{Z})$

$\mathrm{NS}(S) = \mathrm{Ker}(H^2(S,\mathbb{Z}) \longrightarrow H^2(S, \mathcal{O}_S))$

Le groupe $\mathrm{Pic}(S)$ apparait comme une extension :

$$0 \longrightarrow \mathrm{Pic}^\circ(S) \longrightarrow \mathrm{Pic}(S) \longrightarrow \mathrm{NS}(S) \longrightarrow 0$$

de deux groupes de nature différente:

- le groupe $\mathrm{Pic}^\circ(S)$ est un groupe divisible; la théorie de Hodge montre que $H^1(S,\mathbb{Z})$ est un réseau dans $H^1(S, \mathcal{O}_S)$, autrement dit $\mathrm{Pic}^\circ(S)$ a une structure naturelle de tore complexe - et même, en fait, de variété abélienne.

- le groupe $\mathrm{NS}(S) \subset H^2(S,\mathbb{Z})$ est un groupe de type fini.

Le cup-produit sur $H^2(S,\mathbb{Z})$ induit sur $\mathrm{NS}(S)$ une forme bilinéaire symétrique à valeurs dans $\mathbb{Z}$, le produit d'intersection; si D et D' sont deux diviseurs, on note $(D\cdot D')$ le produit de leurs classes dans $\mathrm{NS}(S)$. On obtient ainsi une forme bilinéaire symétrique sur le groupe des diviseurs qui joue un rôle fondamental dans la théorie des surfaces. Si C, C' sont deux courbes irréductibles distinctes, on a :

$(C.C')$ = nombre de points d'intersection de C et C', comptés avec leur multiplicité.

Rappelons quelques théorèmes fondamentaux :.

Théorème de Riemann-Roch : $\chi(\mathcal{O}_S(D)) = \chi(\mathcal{O}_S) + 1/2\,(D^2 - D.K)$

Dualité de Serre : $h^i(K-D) = h^{2-i}(D)$, $0 \leq i \leq 2$.

On utilisera très souvent Riemann-Roch sous la forme suivante, qui utilise la dualité de Serre : $h^o(D) + h^o(K-D) \geq \chi(\mathcal{O}_S) + 1/2\,(D^2 - D.K)$

Formule du genre : Soit C une courbe irréductible sur S .

On a : $\dim H^1(C,\mathcal{O}_C) = 1 + 1/2(C^2 + C.K)$

Ce nombre est le genre de C , noté $g(C)$. Si C est singulière, son genre est strictement plus grand que celui de sa normalisée; en particulier, on a $g(C) = 0$ si et seulement si $C \cong \mathbb{P}^1$.

Invariants numériques

On pose :

$q(S) = \dim H^1(S, \mathcal{O}_S) = \dim H^o(S, \Omega_S^1)$ (par théorie de Hodge)

$p_g(S) = \dim H^2(S, \mathcal{O}_S) = \dim H^o(S, \Omega_S^2) = \dim H^o(K)$ (par dualité de Serre)

$P_n(S) = \dim H^o(nK)$ pour $n \geq 1$.

On notera simplement q, p_g, P_n s'il n'y a pas de confusion possible. Tous ces invariants sont des invariants birationnels. On a :

$$\chi(\mathcal{O}_S) = 1 - q + p_g$$

On pose $b_i = \dim_{\mathbb{C}} H^i(X,\mathbb{C})$; par dualité de Poincaré, on a $b_4 = b_o = 1$ et $b_3 = b_1$. De plus, il résulte de la théorie de Hodge que $b_1 = 2q$.

On pose $\chi_{top}(S) = \sum (-1)^i\, b_i = 2 - 2b_1 + b_2$

Ces invariants sont reliés par la :

Formule de M. Noether : $12\chi(\mathcal{O}_S) = K^2 + \chi_{top}(S)$

Nous utiliserons la variété d'Albanese d'une surface S ; rappelons ici les propriétés qui nous intéressent :

Rappel : variété d'Albanese.

Il existe une variété abélienne $Alb(S)$, de dimension q, et un morphisme $\alpha : S \longrightarrow Alb(S)$ tels que :

- Si $q \geq 1$, $\alpha(S)$ n'est pas réduit à un point;
- Si $\alpha(S)$ est une courbe B, cette courbe est lisse de genre q et la fibration $p : S \longrightarrow B$ a ses fibres connexes.

  On dira que p est la fibration d'Albanese de S.

Nous utiliserons également le théorème classique suivant :

"Théorème de Bertini" : Soient S une surface, C une courbe, $q : S \longrightarrow C$ un morphisme surjectif. Il existe une courbe lisse B et un diagramme commutatif:

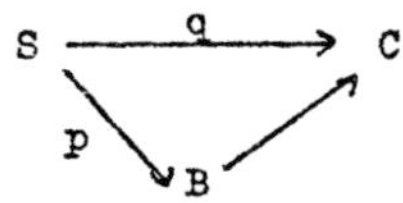

tel que le morphisme $p : S \longrightarrow B$ ait ses fibres connexes.

Notons que la fibre générique du morphisme p est alors lisse et irréductible, puisque la fibre générique d'un morphisme de variétés lisses sur $\mathbb{C}$ est toujours lisse (théorème de Sard) .

Enfin la remarque suivante est triviale mais extrêmement utile :

Remarque utile :

Soient C une courbe irréductible sur S, telle que $C^2 \geqslant 0$, D un diviseur effectif. Alors $(D.C) \geqslant 0$.

Démonstration : on écrit $D = D' + nC$, où D' ne contient pas C, et $n \geqslant 0$; alors $(D.C) = (D'.C) + n(C^2) \geqslant 0$.

§2. Applications birationnelles.

On peut classifier les surfaces à isomorphisme près, ou, plus grossièrement, à isomorphisme birationnel près. Le problème ne se pose pas pour les courbes, puisque toute application birationnelle d'une courbe lisse dans une autre est partout définie. Pour les surfaces, on va voir que toute application birationnelle s'obtient à partir de transformations "élémentaires", les éclatements.

Rappel : Soit S une surface, $p \in S$. Il existe une surface $\hat{S}$ et un morphisme birationnel $\varepsilon : \hat{S} \rightarrow S$, tels que :

- ε restreint à $\varepsilon^{-1}(S-p)$ est un isomorphisme sur $S-p$;
- $\varepsilon^{-1}(p) = E$ est une courbe isomorphe à $\mathbb{P}^1$, qui s'identifie naturellement à l'ensemble des directions tangentes à S en p.

On dit que ε est l'éclatement de S en p, et E la droite exceptionnelle de l'éclatement. On a :

$$\mathrm{Pic}(\hat{S}) \cong \mathrm{Pic}(S) \oplus \mathbb{Z}[E]$$

$$NS(\hat{S}) \cong NS(S) \oplus \mathbb{Z}[E] \qquad \text{(b)}$$

La forme d'intersection sur $\hat{S}$ étant donnée par les formules :

$$(\varepsilon^* D \,.\, \varepsilon^* D') = (D.D') \qquad D, D' \text{ diviseurs sur } S$$

$$(\varepsilon^* D.E) = 0$$

$$E^2 = -1$$

De plus on a :

$K_{\hat{S}} = \varepsilon^* K_S + E$

Soit C une courbe irréductible sur S, passant par p avec multiplicité m. On définit le transformé strict $\hat{C}$ de C comme l'adhérence dans $\hat{S}$ de $\varepsilon^{-1}(C-p)$. On vérifie immédiatement que :

$$\varepsilon^* C = \hat{C} + mE$$

d'où l'on déduit : $\hat{C}^2 = C^2 - m^2$ et

$$\hat{C}.\hat{K} = C.K + m \qquad (c)$$

Nous admettons sans démonstration les théorèmes suivants (cf. par exemple [Ch. 1]) :

Théorème d'élimination des indéterminations.

Soient S *une surface,* V *une variété algébrique,* $\varphi : S \dashrightarrow V$ *une application rationnelle. Il existe un morphisme* $\eta : \tilde{S} \longrightarrow S$, *composé d'une suite finie d'éclatements, et un morphisme* $f : \tilde{S} \longrightarrow V$ *tel que le diagramme :*

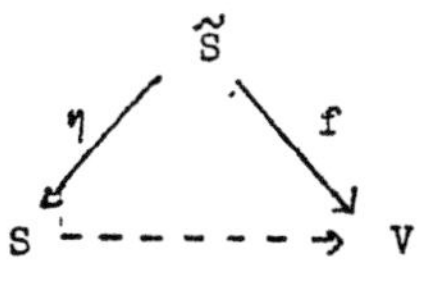

soit commutatif.

Théorème de structure des morphismes birationnels.

Tout morphisme birationnel (d'une surface dans une autre) est composé d'une suite finie d'éclatements.

Corollaire

Soit $\varphi : S \dashrightarrow S'$ *une application birationnelle. Il existe une surface* S *et un diagramme commutatif :*

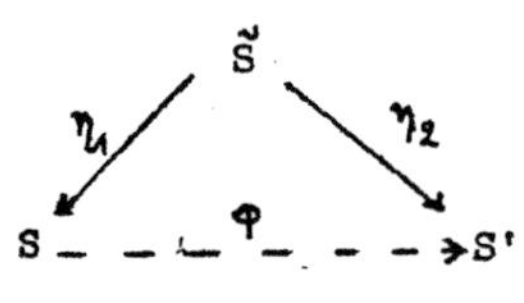

<u>où η_1, η_2 sont des composés d'un nombre fini d'éclatements.</u>

Soit $f : S \rightarrow S'$ un morphisme birationnel. Le nombre n d'éclatements dont il est composé est déterminé par la formule $NS(S) \cong NS(S') \oplus \mathbb{Z}^n$; en particulier, tout morphisme birationnel de S dans elle-même est un isomorphisme.

Pour toute surface S, notons $B(S)$ l'ensemble des classes d'isomorphisme de surfaces birationnellement isomorphes à S. Si S_1, $S_2 \in B(S)$, on dit que S_1 domine S_2 s'il existe un morphisme birationnel (i.e. un composé d'éclatements) de S_1 dans S_2. D'après ce qui précède, on introduit ainsi une relation d'ordre sur $B(S)$. On dit qu'une surface S est minimale si elle est minimale dans $B(S)$, c'est-à-dire si tout morphisme birationnel de S dans une surface S' est un isomorphisme.

<u>Proposition</u>

<u>Toute surface domine une surface minimale.</u>

<u>Démonstration</u> : Soit S une surface. Si S n'est pas minimale, il existe un morphisme birationnel $S \rightarrow S_1$ qui n'est pas un isomorphisme. Si S_1 n'est pas minimale, il existe de même $S \rightarrow S_2$, et ainsi de suite; comme :

$$\text{rg } NS(S) > \text{rg } NS(S_1) > \text{rg } NS(S_2) > \dots \qquad \text{(formule (b) p. 6)}$$

on arrive nécessairement à une surface minimale dominée par S.

Disons qu'une courbe $E \subset S$ est exceptionnelle s'il existe un éclatement $\varepsilon : S \longrightarrow S'$ (S' surface lisse) tel que E soit la droite exceptionnelle de ε ; il résulte du théorème de structure des morphismes birationnels <u>qu'une surface est minimale si et seulement si elle ne contient pas de</u>

courbe exceptionnelle.

Les courbes exceptionnelles sont caractérisées par le théorème suivant que nous admettrons :

Critère de contraction de Castelnuovo

Une courbe E est exceptionnelle si et seulement si $E \simeq \mathbb{P}^1$ et $E^2 = -1$

L'ensemble $B(S)$ sera en principe connu dès que l'on connaîtra ses éléments minimaux - tous les autres étant obtenus à partir de ceux-là par des éclatements. Deux cas peuvent se présenter : il y a un seul modèle minimal, ou il y en a plusieurs.

Définition 1 : Une surface S est réglée si elle est birationnellement isomorphe à $C \times \mathbb{P}^1$, où C est une courbe lisse. Si de plus $C = \mathbb{P}^1$, on dit que S est rationnelle.

Théorème des modèles minimaux

Soient S, S' deux surfaces minimales, $\varphi : S \dashrightarrow S'$ une application birationnelle. Si S n'est pas réglée, φ est un isomorphisme.

En particulier, $B(S)$ a un seul élément minimal, et tout automorphisme birationnel de S est un automorphisme.

La démonstration utilise le lemme fondamental de la théorie des surfaces:

Lemme-clé (Enriques-Castelnuovo)

Soient S une surface minimale non réglée, C une courbe irréductible.

Alors : $K.C \geqslant 0$.

Ce lemme sera démontré plus tard (§8) .

Démonstration du théorème : Par le théorème d'élimination des indéterminations, il existe un diagramme commutatif :

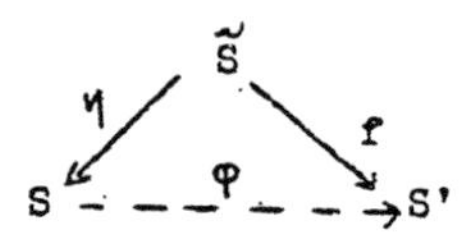

où $\eta = \varepsilon_n \circ \ldots \circ \varepsilon_1$, est un composé de n éclatements et f un morphisme birationnel.

Parmi tous les diagrammes possibles, choisissons-en un tel que n soit minimal; il s'agit de montrer que $n = 0$. Soit E la droite exceptionnelle de l'éclatement ε_1 . Si $f(E)$ était réduit à un point, le morphisme f se factoriserait en $f' \circ \varepsilon_1$ et l'on contredirait la minimalité de n . Donc $f(E)$ est une courbe C . Comme f est un composé d'éclatements, il résulte de la formule (c) p.7 que :

$$(C.K_{S'}) \leq (E.K_{\tilde{S}}) = -1$$

d'où une contradiction avec le lemme-clé.

La classification des surfaces se divise donc en deux branches : d'un côté les surfaces réglées, qu'on peut considérer comme connues du point de vue birationnel, mais dont on cherchera les modèles minimaux; de l'autre les surfaces non réglées, pour lesquelles la classification "birégulière" revient essentiellement au même que la classification birationnelle : il suffira de classer les surfaces minimales.

§3. Surfaces réglées et rationnelles

Définition 2

Soit C une courbe lisse. Une surface géométriquement réglée de base C est une surface S , munie d'un morphisme lisse $p : S \to C$ dont les fibres sont isomorphes à $\mathbb{P}^1$.

Il n'est pas évident a priori qu'une surface géométriquement réglée est réglée (Déf.1); cela résulte du :

Théorème 3 (Noether-Enriques)

Soient S une surface, p un morphisme de S sur une courbe lisse C. On suppose qu'il existe $x \in C$ tel que la fibre $p^{-1}(x)$ soit isomorphe à $\mathbb{P}^1$. Alors il existe un ouvert U de C, de la forme $U = C - \{x_1 \ldots x_n\}$ $(x_i \neq x)$ et un isomorphisme de $p^{-1}(U)$ sur $U \times \mathbb{P}^1$, commutant avec les projections sur U.

Pas 1 : $p_g(S) = 0$

Notons $F = p^{-1}(x)$. On a $F^2 = 0$ et $F.K = -2$ (formule du genre), donc si $D \in |K|$ on doit avoir $D.F = -2$ mais aussi $D.F \geqslant 0$ par la remarque utile (p. 6) ,et par suite $|K| = \emptyset$.

Pas 2 : il existe un diviseur H sur S tel que $(H.F) = 1$.

Comme $p_g(S) = 0$, la flèche $\mathrm{Pic}(S) \longrightarrow H^2(S,\mathbb{Z})$ est surjective ((a)p. 3). Il suffit donc de montrer qu'il existe une classe $h \in H^2(S,\mathbb{Z})$ telle que $h.f = 1$, en notant f la classe de F dans $H^2(S,\mathbb{Z})$. Pour a variable dans $H^2(S,\mathbb{Z})$, l'ensemble des entiers $(a.f)$ est un idéal de $\mathbb{Z}$, de la forme $d.\mathbb{Z}$ $(d \geqslant 1)$. L'application $a \longmapsto 1/d\,(a.f)$ est une forme linéaire sur $H^2(S,\mathbb{Z})$; par dualité de Poincaré, il existe un élément $f' \in H^2(S,\mathbb{Z})$ tel que :

$$(a.f') = 1/d\,(a.f) \qquad \text{pour tout} \quad a \in H^2(S,\mathbb{Z})$$

et donc $f = d.f'$ modulo torsion dans $H^2(S,\mathbb{Z})$.

Mais comme $f^2 = 0$, $f.[K] = -2$ et que $f'^2 + f'.[K]$ est pair, on voit qu'on a nécessairement $d = 1$, d'où le résultat.

Pas 3

Considérons la suite exacte :

$$0 \longrightarrow \mathcal{O}_S(H + (r-1)F) \longrightarrow \mathcal{O}_S(H+rF) \longrightarrow \mathcal{O}_F(1) \longrightarrow 0 \qquad (r \in \mathbb{Z})$$

On en déduit la suite exacte longue de cohomologie :

$$H^0(S, \mathcal{O}_S(H+rF)) \xrightarrow{a_r} H^0(F, \mathcal{O}_F(1)) \longrightarrow H^1(S, \mathcal{O}_S(H+(r-1)F)) \xrightarrow{b_r} H^1(S, \mathcal{O}_S(H+rF)) \to 0$$

La suite des espaces quotients $H^1(S, \mathcal{O}_S(H+rF)$ doit être stationnaire pour r assez grand; donc il existe un r tel que b_r soit bijectif et par suite a_r surjectif. Choisissons un sous-espace vectoriel V de $H^0(S, \mathcal{O}_S(H+rF))$, de dimension 2, tel que $a_r(V) = H^0(F, \mathcal{O}_F(1))$; notons P le pinceau correspondant. Il peut avoir des composantes fixes, mais elles doivent être contenues dans certaines fibres $F_{x_1}, \dots, F_{x_k}$ de p distinctes de F (puisque P n'a pas de points fixes sur F). De même les points fixes de la partie mobile de P sont contenus dans des fibres $F_{x_{k+1}}, \dots, F_{x_l}$ distinctes de F. Enfin notons $F_{x_{l+1}}, \dots, F_{x_m}$ les fibres de p qui ne sont pas irréductibles. Posons $U = C - \{x_1, \dots, x_m\}$, et notons P' la restriction de P à $p^{-1}(U)$. Le pinceau P' est sans points fixes; toute courbe C_t de P' est réunion d'une section de p et éventuellement de certaines fibres; mais en fait C_t ne contient pas de fibres, sans quoi on aurait $C_t \cap C_{t'} \neq \emptyset$ pour $t \neq t'$. Donc le pinceau P' est formé de sections $(C_t)_{t \in \mathbb{P}^1}$ de la fibration p. Comme il est sans points fixes, il définit un morphisme $g : p^{-1}(U) \to \mathbb{P}^1$, de fibre $g^{-1}(t) = C_t$; on en déduit un morphisme $h = (p,g)$ de $p^{-1}(U)$ sur $U \times \mathbb{P}^1$. Comme $h^{-1}((x,t)) = F_x \cap C_t$, h est un isomorphisme, d'où le théorème.

Remarque 4

Lorsque p est lisse, S est donc un fibré en droites projectives au-dessus de C, localement trivial (pour la topologie de Zariski). L'ensemble des classes d'isomorphisme de tels fibrés s'identifie à l'ensemble $H^1(C, PGL(2, \mathcal{O}_C))$. Or on déduit de la suite exacte :

$$1 \longrightarrow \mathcal{O}_C^* \longrightarrow GL(2, \mathcal{O}_C) \longrightarrow PGL(2, \mathcal{O}_C) \longrightarrow 1$$

la suite exacte de cohomologie :

$$\mathrm{Pic}(C) \longrightarrow H^1(C, GL(2, \mathcal{O}_C)) \longrightarrow H^1(C, PGL(2, \mathcal{O}_C)) \longrightarrow H^2(C, \mathcal{O}_C^*)$$

Or $H^2(C, \mathcal{O}_C^*) = 0$, par exemple parce que C est une courbe; donc tout fibré en droites projectives sur C est le fibré projectif $\mathbb{P}_C(E)$ associé à un fibré vectoriel E de rang 2 sur C . Les fibrés $\mathbb{P}_C(E)$ et $\mathbb{P}_C(E')$ sont isomorphes si et seulement si il existe un fibré inversible L sur C tel que $E' \cong E \otimes L$.

La classification des surfaces géométriquement réglées sur C est donc ramenée à celle des fibrés vectoriels de rang 2 sur C . Celle-ci est loin d'être triviale, mais peut être considérée comme bien comprise - cf. [R] .

Lemme 5

Soient S une surface, C une courbe lisse, $p : S \longrightarrow C$ un morphisme dont la fibre générique est isomorphe à $\mathbb{P}^1$. Si une fibre de p n'est pas irréductible, elle contient une droite exceptionnelle.

Démonstration : Soit F une fibre réductible, $F = \sum n_i C_i$. On a $C_i^2 < 0$ pour tout i (car $n_i C_i^2 = C_i (F - \sum_{j \neq i} n_j C_j) < 0$), donc par la formule du genre $K.C_i \geq -1$, l'égalité n'ayant lieu que si C_i est une courbe exceptionnelle. Par suite si F ne contenait pas de courbes exceptionnelles, on trouverait $K.F \geq 0$, ce qui contredirait $K.F = -2$.

Théorème 6

Soit C une courbe lisse non rationnelle. Les modèles minimaux de $C \times \mathbb{P}^1$ sont les surfaces géométriquement réglées de base C .

Démonstration : Il est clair qu'une surface géométriquement réglée ne contient pas de droites exceptionnelles, car celles-ci devraient s'envoyer surjectivement sur C , ce qui est impossible. Soient S une surface minimale, f une application birationnelle de S sur $C \times \mathbb{P}^1$, p la projection de $C \times \mathbb{P}^1$ sur C . Considérons l'application rationnelle $p \circ f : S \dashrightarrow C$; par le théorème d'élimination des indéterminations il existe un diagramme commutatif :

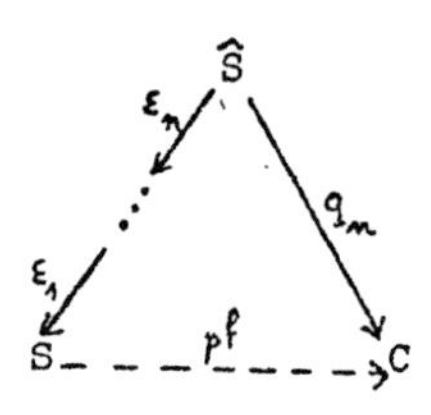

où les ε_i sont des éclatements, et q_n un morphisme. Notons E_n la droite exceptionnelle de l'éclatement ε_n . Comme C n'est pas rationnelle, $q_n(E_n)$ est réduit à un point, de sorte que q_n se factorise en $q_{n-1} \circ \varepsilon_n$. En continuant le procédé, on voit finalement que pf est un morphisme, de fibre générique isomorphe à $\mathbb{P}^1$.

D'après le lemme 5, les fibres de pf sont irréductibles, donc rationnelles lisses (puisqu'elles vérifient $F^2 = 0$, $F.K = -2$, d'où $g(F) = 0$) : par suite S est une surface géométriquement réglée de base C .

Nous nous contenterons d'énoncer sans démonstration le résultat analogue pour les surfaces rationnelles. On sait que tout fibré vectoriel sur $\mathbb{P}^1$ est somme de fibrés inversibles; il en résulte que les surfaces géométriquement réglées de base $\mathbb{P}^1$ sont les surfaces :

$$F_n = \mathbb{P}_{\mathbb{P}^1}(\mathcal{O}_{\mathbb{P}^1} \oplus \mathcal{O}_{\mathbb{P}^1}(n)) \qquad \text{pour } n \geqslant 0 \quad .$$

<u>Théorème</u>

<u>Les surfaces rationnelles minimales sont</u> $\mathbb{P}^2$ <u>et les surfaces</u> F_n <u>pour</u> $n \neq 1$.

Remarque 7

Il est facile de calculer les invariants numériques des surfaces réglées : si S est birationnellement équivalente à $C \times \mathbb{P}^1$, on trouve :

- $P_n(S) = 0$ pour tout n

- $q(S) = g(C)$.

Si de plus S est minimale $\neq \mathbb{P}^2$, on a :

$$K_S^2 = 8(1-q) \qquad b_2(S) = 2$$

tandis que $K^2_{\mathbb{P}^2} = 9$ et $b_2(\mathbb{P}^2) = 1$.

Nous allons voir qu'inversement l'annulation des P_n caractérise les surfaces réglées.

§4. Caractérisation des surfaces rationnelles et réglées.

Rappelons que nous démontrerons plus loin (§8) le :

Lemme-clé : *Soit S une surface minimale non reglée, D un diviseur effectif sur S. Alors $D.K \geqslant 0$.*

Lemme 8

Soit S une surface minimale avec $K^2 < 0$. Alors S est réglée.

Démonstration : Soit H une section hyperplane de S. Si $(H.K) < 0$, on applique le lemme-clé; de même si $(H.K) = 0$, en prenant $D = nH+K$ qui est effectif pour n assez grand. On peut donc supposer $(H.K) > 0$.

Posons $r_o = \frac{H.K}{-K^2}$. On a :

$(H+r_o K)^2 = H^2 + \frac{(H.K)^2}{-K^2} > 0$ et $(H+r_oK).K = 0$, de sorte que si r

est un rationnel $> r_o$ et suffisament voisin de r_o, on a :

$$(H+rK)^2 > 0 \qquad (H+rK).K < 0 \qquad (H+rK).H > 0$$

Posons $D_m = m(H+rK)$, pour tout m tel que $mr \in \mathbb{Z}$. Par Riemann-Roch, on a : $h^o(D_m) + h^o(K-D_m) \longrightarrow \infty$ quand $m \longrightarrow \infty$.

Comme $(K-D_m).H$ devient négatif pour m grand, on voit que pour m grand le système $|D_m|$ est non vide; comme $D_m.K < 0$, on conclut encore par le lemme-clé.

Théorème 9 (Castelnuovo)

Soit S une surface avec $q = P_2 = 0$. Alors S est rationnelle.

Démonstration : On peut supposer S minimale. Si $K^2 < 0$, on applique le lemme 8. Si $K^2 \geq 0$, le théorème de Riemann-Roch (compte tenu de $P_2 = 0$) donne : $h^o(-K) \geq 1 + K^2 \geq 1$ donc si H est une section hyperplane, on a $(K.H) < 0$, d'où le résultat par le lemme-clé.

Notre but est maintenant de caractériser les surfaces réglées par l'annulation des plurigenres P_n. Si $q = 0$, l'annulation de P_2 suffit; Si $q \geq 1$, celle de p_g n'est pas loin de suffire :

Lemme 10

Soit S une surface minimale avec $p_g = 0$, $q \geq 1$. On a alors $K^2 < 0$ (et donc S est réglée) sauf si $q = 1$; $b_2 = 2$, $K^2 = 0$.

Démonstration : la formule de Noether (p. 5) s'écrit ici :

$$10 - 8q = K^2 + b_2$$

Il suffit donc de vérifier qu'on ne peut avoir $q = 1$, $b_2 = 1$. On considère pour cela la fibration d'Albanese $p : S \longrightarrow B$, où B (= Alb(S)) est une courbe elliptique; il est clair qu'une section hyperplane de S et une fibre générique de p sont linéairement indépendantes dans $NS(S)$, de sorte que $b_2 \geq 2$.

Proposition 11

Soit S une surface minimale non réglée avec $p_g = 0$, $q = 1$. Alors $S = (C \times F)/G$, où C et F sont des courbes lisses de genre ≥ 1, G un groupe fini d'automorphismes de C, opérant sans points fixes sur $C \times F$ (de manière compatible avec la projection sur C), C/G est elliptique, et C ou F est elliptique.

Démonstration (rapide): On considère la fibration d'Albanese $p : S \longrightarrow B$, où $B = \mathrm{Alb}(S)$ est une courbe elliptique. On désigne par F_b la fibre $p^{-1}(b)$ pour $b \in B$, et par F_η une fibre générique de p. On pose $g = g(F_\eta)$.

Pas 1 : Si $g \geq 2$, p est lisse; si $g = 1$, les fibres de p sont soit lisses, soit de la forme nE, où E est une courbe elliptique lisse.

En premier lieu le fait que $b_2 = 2$ entraîne que les fibres sont irréductibles. On utilise ensuite la formule topologique suivante :

$$\chi_{top}(S) = \chi_{top}(B) \cdot \chi_{top}(F_\eta) + \sum_{b \in B} (\chi_{top}(F_b) - \chi_{top}(F_\eta)) \qquad (d)$$

Lorsqu'une fibre générique F_η se spécialise en une fibre singulière (irréductible) F_b, un certain nombre de 1-cycles sur F (les "cycles évanescents") disparaissent dans l'homologie de F_b ; autrement dit, on a $b_1(F_b) < b_1(F_\eta)$, et donc $\chi_{top}(F_b) > \chi_{top}(F_\eta)$. Or par hypothèse $\chi_{top}(S) = \chi_{top}(B) = 0$; la formule (d) montre donc qu'il ne peut y avoir de fibres singulières. Enfin si $F_b = nC$, on trouve :

$$\chi_{top}(F_b) = \chi_{top}(C) = 2\chi(\mathcal{O}_C) = (C.K) = \frac{1}{n}(F_b.K) = \frac{1}{n}(F_\eta.K) = \frac{1}{n}\chi_{top}(F_\eta)$$

donc $\chi_{top}(F_b) > \chi_{top}(F_\eta)$ sauf si $\chi_{top}(F_b) = \chi_{top}(F_\eta) = 0$.

Pas 2 : Si p est lisse, il existe un revêtement étale $C \longrightarrow B$ tel que la fibration image réciproque $\tilde{S} = S \times_B C \longrightarrow C$ soit triviale (i.e. $\tilde{S} \cong C \times F_\eta$).

La fibration p définit une famille de courbes de genre g sur B. Quitte à passer à un revêtement étale C de B, on peut "rigidifier" la cohomologie (mod n) de ces courbes, c'est-à-dire rendre constant le système localement constant des $H^1(F_b, \mathbb{Z}/n\mathbb{Z})$ et en choisir une base symplectique. Si $n \geq 3$, on sait ([G]) qu'il existe une famille de courbes de genre g $P : U_{n,g} \longrightarrow T_{n,g}$, à cohomologie (mod n) rigidifiée, qui est universelle; c'est-à-dire qu'il existe un morphisme $f : C \longrightarrow T_{n,g}$ telle que la fibration $\tilde{S} \longrightarrow C$ se déduise de P par image réciproque. Or le revêtement universel de C est $\mathbb{C}$ et celui de $T_{n,g}$ est l'espace de Teichmüller T_g qui est un domaine borné, donc le morphisme f est trivial (i.e. $f(C)$ est un point), ce qui implique que la fibration $\tilde{S} \longrightarrow C$ est triviale. On peut supposer le revêtement $C \longrightarrow B$ galoisien de groupe G, de sorte que $S = (C \times F)/G$, avec C elliptique.

(Au lieu d'utiliser le théorème difficile de structure de l'espace de Teichmüller, on peut considérer l'espace $\mathfrak{A}_{n,g}$ des modules des variétés abéliennes principalement polarisées, à cohomologie (mod n) rigidifiée; son revêtement universel est l'espace de Siegel H_g, qui est un domaine borné pratiquement par construction. Il faut alors utiliser le théorème de Torelli.)

Pas 3 : Si p a des fibres multiples (donc $g = 1$), il existe un revêtement ramifié $C \longrightarrow B$ tel que si $\tilde{S}$ désigne la normalisée de $S \times_B C$, la fibration $\tilde{S} \longrightarrow C$ déduite de p soit triviale.

Soit B' une courbe lisse sur S, telle que $p(B') = B$; notons S' la normalisée de $S \times_B B'$. Alors S' est lisse; on montre que pour un choix convenable de B', la projection $S' \longrightarrow S$ est étale. La fibration $p' : S' \longrightarrow B'$ déduite de p possède une section : elle n'a donc pas de fibres multiples, autrement dit p' est lisse. Quitte à passer à un revêtement étale C de B', on a alors comme précédemment un morphisme $f : C \longrightarrow T_{n,1} (n \geq 3)$. Or $T_{n,1}$ est une courbe affine (c'est un revêtement ramifié de $\mathbb{C}$, via l'invariant j), donc f est trivial et la fibration sur C déduite de p' est triviale. On peut supposer le revêtement $C \longrightarrow B$ galoisien de groupe G, de sorte que $S = (C \times F_\eta)/G$, avec

F_η elliptique.

La proposition est donc démontrée.

Corollaire 12

Soit S *une surface minimale non réglée avec* $p_g = 0$, $q \geq 1$.

(i) *Il existe un entier* $n \geq 0$ *tel que* $P_n \neq 0$;

(ii) *Si* $S = (E \times F)/G$, *où* E *et* F *sont elliptiques, on a* $nK \equiv 0$;

(iii) *Si* S *n'est pas du type précédent, les* P_m *ne sont pas bornés.*

Démonstration : Il résulte de la proposition qu'il existe un revêtement étale de degré n $\pi : C \times F \longrightarrow S$, avec $g(C)$,$g(F) \geq 1$. On a $p_g(C \times F) = g(C).g(F) \geq 1$, donc il existe un diviseur effectif $D \in |K_{C \times F}|$; comme π est étale, on sait que $K_{C \times F} \equiv \pi^* K_S$, de sorte que :

$$\pi_* D \in |\pi_* \pi^* K_S| = |n.K_S| \quad \text{d'où} \quad P_n(S) \geq 1 .$$

Le même argument montre que $P_{rn}(S) \geq P_r(C \times F)$, ce qui montre que les P_n sont non bornés dès que $g(C)$ ou $g(F) \geq 2$. Enfin si C et F sont elliptiques, on a $K_{C \times F} \equiv 0$; donc si $D \in |nK_S|$, le diviseur $\pi^* D$ est nul, ce qui signifie que $D = 0$, i.e. $n.K_S \equiv 0$.

Du théorème de Castelnuovo, du lemme 10 et du corollaire 12 résulte le :

Théorème 13 (Enriques)

Une surface S *est réglée si et seulement si* $P_n = 0$ *pour tout* $n \geq 0$.

Remarque 14

Une étude plus approfondie des surfaces du type $(C \times F)/G$ donne la condition d'Enriques : S est réglée si et seulement si $P_{12} = 0$.

Les surfaces du type $S = (E \times F)/G$, où E et F sont elliptiques et $p_g = 0$, sont appelées surfaces bielliptiques (ou parfois hyperelliptiques, mais cette terminologie prête à confusion). On peut en donner une classification complète.

§5. Dimension de Kodaira

Les résultats qui précèdent montrent l'importance des P_n dans la classification des surfaces. Ils conduisent à poser la :

Définition 15

Soit S une surface; pour $n \geqslant 0$, notons φ_{nK} l'application rationnelle de S dans un espace projectif (éventuellement vide) définie par le système $|nK|$. La "dimension de Kodaira" de S, notée $\kappa(S)$ (ou simplement κ) est la plus grande dimension des images des φ_{nK} pour $n \geqslant 0$.

(On convient de poser $\dim(\emptyset) = -1$).

Explicitons la définition :

- $\kappa(S) = -1 \iff P_n = 0$ pour tout $n \iff S$ réglée (par le théorème d'Enriques).
- $\kappa(S) = 0 \iff P_n = 0$ ou 1, et il existe N tel que $P_N = 1$.
- $\kappa(S) = 1 \iff$ Il existe N tel que $P_N \geqslant 2$; et pour tout n, l'image de φ_{nK} est au plus une courbe.
- $\kappa(S) = 2 \iff$ Il existe N tel que l'image de φ_{NK} soit une surface.

Une définition analogue peut être donnée pour les courbes; on voit aussitôt que $\kappa(\mathbb{P}^1) = -1$, $\kappa(C) = 0$ si et seulement si C est elliptique, et $\kappa(C) = 1$ si et seulement si $g(C) \geqslant 2$.

Exemple 16

1/ $S = C \times C'$. On vérifie immédiatement que :

- si C ou $C' = \mathbb{P}^1$, $\kappa(S) = -1$;
- si C et C' sont elliptiques, $\kappa(S) = 0$;
- si C est elliptique, et $g(C') \geqslant 2$, $\kappa(S) = 1$;
- si $g(C)$ et $g(C') \geqslant 2$, $\kappa(S) = 2$.

2/ $S = V_{d_1,\ldots,d_r}$ = intersection complète dans $\mathbb{P}^{r+2}$ de r hypersurfaces de degrés $d_1, \ldots, d_r$.

Un calcul facile montre que $K_S \equiv (\sum d_i - r - 3)H$, où H est une section hyperplane de S . Il en résulte que :

$\kappa(S) = -1$ pour $S = V_2$, V_3 , $V_{2,2}$

$\kappa(S) = 0$ (et en fait $K_S \equiv 0$) pour $S = V_4$, $V_{2,3}$, $V_{2,2,2}$

$\kappa(S) = 2$ pour les autres.

3/ Toute surface telle que $n \cdot K \equiv 0$ pour un entier n , en particulier toute surface bielliptique (remarque 14) ou abélienne, a dimension de Kodaira zéro.

§6. Surfaces avec $\kappa = 0$.

Ce sont les surfaces avec $P_n = 0$ ou 1 pour tout n , et
$P_n = 1$ pour au moins un n .

Lemme 17

Soit S *minimale avec* $\kappa = 0$.

a/ *On a* $K^2 \geqslant 0$

b/ *On a* $\chi(\mathcal{O}_S) \geqslant 0$

c/ *Si* $P_n = P_m = 1$, *et* $d = (n,m)$, *on a* $P_d = 1$.

Démonstration : a/ On a $K^2 \geqslant 0$ par le lemme 8 ; supposons $K^2 > 0$. On a par Riemann-Roch :

$$h^0(nK) + h^0((1-n)K) \longrightarrow \infty \quad \text{quand} \quad n \longrightarrow \infty .$$

Pour $n \geqslant 2$, le système $|(1-n)K|$ ne peut contenir un diviseur E , sans quoi on aurait $(E.K) \geq 0$ (Lemme-clé) et donc $K^2 \leq 0$; on trouve donc que $P_n \longrightarrow \infty$ quand $n \to \infty$, ce qui contredit $\kappa = 0$.

b/ Comme $K^2 = 0$, la formule de Noether s'écrit :

$$12\ \chi(\Theta_S) = \chi_{top}(S) = 2 - 4q + b_2$$

soit : $8\chi(\Theta_S) = -2 - 4p_g + b_2 \geq -6$ (puisque $p_g \leq 1$)

d'où le résultat.

c/ Soient $D \in |nK|$, $E \in |mK|$; posons $m = m'd$, $n = n'd$. Comme $P_{\frac{nm}{d}} = 1$, on a $m'D = n'E \in |\frac{mn}{d} K|$, d'où $D = n'\Delta$, $E = m'\Delta$ pour un diviseur effectif Δ. Posons $\varepsilon = \Delta - dK$ dans $\mathrm{Pic}(S)$; on a $m'\varepsilon = n'\varepsilon = 0$, d'où $\varepsilon = 0$ puisque $(m',n') = 1$; donc $\Delta \in |dK|$ et $P_d = 1$.

Théorème 18

Soit S une surface minimale avec $\kappa = 0$. Une des 4 possibilités suivantes est réalisée :

1/ $p_g = 0$, $q = 0$. Alors $2K \equiv 0$. On dit que S est une "surface d'Enriques".

2/ $p_g = 0$, $q = 1$: S est une surface bielliptique (remarque 14)

3/ $p_g = 1$, $q = 0$. Alors $K \equiv 0$. On dit que S est une "surface K3".

4/ $p_g = 1$, $q = 2$. Alors S est une surface abélienne.

Démonstration : 1/ Si $p_g = 0$, $q = 0$, on a $P_2 \geq 1$ par le théorème de Castelnuovo, d'où par Riemann-Roch :
$h^0(-2K) + h^0(3K) \geq 1$.

Comme $p_g = 0$, on doit avoir $P_3 = 0$ par le lemme 17. c/, donc $h^0(-2K) \geq 1$; par suite $2K \equiv 0$.

2/ Les surfaces minimales avec $p_g = 0$, $q \geq 1$ ont été classifiées (§4); il résulte du corollaire 12 que celles qui vérifient $\kappa = 0$ sont les surfaces biellipliques.

— Supposons maintenant $p_g = 1$. Par le lemme 17 b/, on a $q = 0$, 1 ou 2.

3/ Si $q = 0$, Riemann-Roch donne $h^0(-K) + h^0(2K) \geq 2$, d'où $h^0(-K) = 1$ et $K \equiv 0$.

3'/ Si $q = 1$, il existe un diviseur ε tel que $\varepsilon \not\equiv 0$ mais $2\varepsilon \equiv 0$. Appliquons-lui Riemann-Roch :

$$h^0(\varepsilon) + h^0(K-\varepsilon) \geq 1 \quad \text{d'où} \quad h^0(K-\varepsilon) \geq 1 .$$

Soit $D \in |K-\varepsilon|$, $K_o \in |K|$; puisque $P_2 = 1$, on a $2D = 2K_o$ et donc $D = K_o$, ce qui contredit $\varepsilon \not\equiv 0$. Il existe donc pas de surface minimale avec $\kappa = 0$, $p_g = 1$, $q = 1$.

4/ Il reste à démontrer qu'une surface vérifiant $\kappa = 0$, $p_g = 1$, $q = 2$ est une surface abélienne. C'est l'objet de la proposition 20.

Lemme 19

Soient S une surface, B une courbe lisse, $p : S \to B$ un morphisme surjectif à fibres connexes, $C_1, \dots, C_r$ les composantes irréductibles d'une fibre F_b de p, $D = \sum n_i C_i$ $(n_i \in \mathbb{Z})$.

Alors $D^2 \leq 0$, et $D^2 = 0$ si et seulement si $D = r.F_b$ $(r \in \mathbb{Q})$.

Démonstration : Posons $F_b = \sum m_i C_i$, $m_i > 0$. On a :

$$D^2 = \sum_i n_i^2 C_i^2 + 2 \sum_{i<j} n_i n_j (C_i.C_j)$$

Eliminons les C_i^2 en utilisant le fait que $(F_b.C_i) = 0$:

$$D^2 = \sum_i n_i^2 (C_i \cdot \sum_{j \neq i} - \frac{m_j}{m_i} \cdot C_j) + 2 \sum_{i<j} n_i n_j (C_i.C_j)$$

$$= \sum_{i<j} (C_i.C_j)\left(- n_i^2 \frac{m_j}{m_i} - n_j^2 \frac{m_i}{m_j} + 2\, n_i\, n_j\right)$$

$$= - \sum_{i<j} (C_i.C_j)\, m_i\, m_j \left(\frac{n_i}{m_i} - \frac{n_j}{m_j}\right)^2 \leq 0$$

Comme $\cup C_i$ est connexe, on n'a égalité que si $\frac{n_i}{m_i} = \frac{n_j}{m_j}$ pour tous i,j, d'où le lemme.

Proposition 20

Soit S une surface minimale avec $\kappa = 0$, $p_g = 1$, $q = 2$. Alors S est une surface abélienne.

Démonstration : Notons $A = \mathrm{Alb}(S)$ et $\alpha : S \to A$ le morphisme d'Albanese. On distinguera quatre cas, suivant que $\alpha(S)$ est une courbe ou une surface, et K_S est trivial ou non.

1/ $K_S \neq 0$

Notons K le diviseur effectif de $|K_S|$. Ecrivons $K = \sum n_i C_i$; comme $K^2 = 0$ et $K.C_i \geqslant 0$ pour tout i, on a $K.C_i = 0$; par suite :
$0 = K.C_i = n_i C_i^2 + \sum_{j \neq i} n_j (C_i.C_j)$
donc ou bien $C_i^2 = -2$ et C_i est rationnelle lisse (car $K.C_i = 0$), ou bien $C_i^2 = 0$, $C_i.C_j = 0$ pour tout $j \neq i$.

On conclut que si l'on écrit $K = \sum_\alpha D_\alpha$, où les D_α sont des diviseurs effectifs à supports connexes et disjoints, on a $D_\alpha^2 = 0$ pour tout α et :

- ou bien D_α est une courbe irréductible de genre 1 (i.e. une courbe elliptique lisse ou une courbe rationnelle avec un point double)
- ou bien D_α est réunion de courbes rationnelles lisses.

1 a/ $K \neq 0$ et $\alpha(S)$ est une courbe.

On considère la fibration d'Albanese $p : S \to B$. Comme B est de genre 2 , les diviseurs D_α sont contenus dans des fibres de p ; par le lemme 19 , on doit avoir $D_\alpha = r_\alpha F_{b_\alpha}$ pour tout α , avec $r_\alpha \in \mathbb{Q}^+$ et $b_\alpha \in B$. Pour n convenable on aura donc :

$$nK \equiv \sum n_\alpha F_{b_\alpha} = p^*(\sum n_\alpha b_\alpha) \quad \text{avec} \quad n_\alpha \in \mathbb{Z}^+$$

dès que n est assez grand on aura $P_n \geq 2$, ce qui contredit $\kappa = 0$.

1 b/ $K \not\equiv 0$ et $\alpha(S) = A$.

Comme A ne contient pas de courbes rationnelles, toute composante D_α de K contenant des courbes rationnelles est contractée sur un point; or par un théorème de Mumford, tout diviseur effectif D tel que $\alpha(D)$ soit un point vérifie $D^2 < 0$. Les seuls D_α possibles sont donc de la forme $n E$, où E est une courbe elliptique lisse, telle que $\alpha(E)$ soit une courbe. On sait qu'à translation près, tout morphisme de E dans A est un morphisme de variétés abéliennes; de sorte qu'en choisissant convenablement l'origine de A on peut supposer que $\alpha(E) = E'$ est une sous-variété abélienne de A. Posons $A' = A/E'$, et considérons la fibration $f : S \to A'$. Par le théorème de Bertini (p. 5) , il existe un diagramme commutatif :

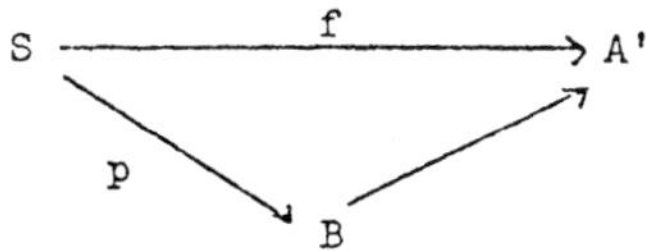

où B est une courbe lisse, et où les fibres de p sont connexes. Comme $p(E)$ est réduit à un point, et $E^2 = 0$, on déduit du lemme 19 que $E = (1/q) F_b$, $q \in \mathbb{Z}$. On conclut alors comme précédemment que $P_n \geq 2$ pour n assez grand.

2 a/ $K \equiv 0$ et $\alpha(S)$ est une courbe.

Posons $\alpha(S) = B$, et choisissons un revêtement étale de degré ≥ 2 $B' \to B$ Soit $S' = S \times_B B'$; S' est connexe, et munie d'un revêtement étale

$\pi : S' \to S$. On a $\chi(\mathcal{O}_{S'}) = 0$ et $K_{S'} \equiv \pi^* K_S \equiv 0$, d'où $p_g(S') = 1$; on en tire $q(S') = 2$. Mais on a $g(B') \geq 3$ et $q(S') \geq g(B')$ (par exemple parce que la flèche $H^o(B', \Omega^1_{B'}) \to H^o(S', \Omega^1_{S'})$ est injective), d'où contradiction.

Il y a donc une seule possibilité :

2 b/ $K \equiv 0$ et $\alpha(S) = A$.

Pour démontrer la proposition, il suffit de montrer que α est étale, puisque tout revêtement étale d'une variété abélienne est lui-même une variété abélienne; autrement dit, si ω est une 2-forme partout $\neq 0$ sur A, il suffit de montrer que $\alpha^* \omega$ est partout $\neq 0$ sur S. Comme α est génériquement étale, la forme $\alpha^* \omega$ n'est pas identiquement nulle; puisque $K_S \equiv 0$, elle est donc partout $\neq 0$.

§7. Surfaces avec $\kappa = 1$ et $\kappa = 2$.

Proposition 21

Soit S une surface minimale. Les conditions suivantes sont équivalentes.

a/ $\kappa = 2$

b/ $K^2 > 0$ et S non rationnelle

c/ Il existe un entier n_o tel que tout $n \geq n_o$, l'application rationnelle $\varphi_{nK} : S \dashrightarrow \mathbb{P}^N$ est birationnelle.

Démonstration : Il est clair que c/ $\Rightarrow$ a/. Montrons que b/ $\Rightarrow$ c/ : soit H une section hyperplane de S. Comme $K^2 > 0$, le théorème de Riemann-Roch donne : $h^o(nK - H) + h^o((1-n)K + H) \to \infty$ quand $n \to \infty$.
Puisque $((1-n)K + H).K$ devient négatif pour n grand, il résulte du lemme-clé que le système $|(1-n)K + H|$ est vide, donc pour $n \geq n_o$ $|nK - H|$ contient un diviseur effectif E_n : on a $nK = H + E_n$. Il est

alors clair que l'application rationnelle φ'_{nK} , restreinte à $S - E_n$, est un plongement.

L'implication a/ $\Rightarrow$ b/ résulte aussitôt du :

Lemme 22

Soit S une surface minimale avec $K^2 = 0$. Supposons que pour un $n \geqslant 1$, le système $|nK|$ ne soit pas réduit à un seul diviseur, de sorte que $nK = Z + M$, où Z est la partie fixe de $|nK|$ et $|M|$ n'a pas de composantes fixes. On a alors $K.Z = K.M = Z^2 = Z.M = M^2 = 0$; le système M n'a pas de points fixes, et définit un morphisme (égal à φ_{nK}) de S sur une courbe.

Démonstration : S est non réglée, donc $K.Z$ et $K.M$ sont $\geqslant 0$ (lemme-clé); comme : $nK^2 = K.Z + K.M$ on a $K.Z = K.M = 0$. Puisque le diviseur M est mobile, on a $M^2 \geqslant 0$ et $M.Z \geqslant 0$; or :

$$M^2 = M(nK - Z) = -M.Z \qquad \text{donc} \qquad M^2 = M.Z = 0$$

et par suite $Z^2 = Z.(nK-M) = 0$.

Comme $M^2 = 0$, le système $|M|$ est sans points fixes, et définit donc un morphisme $\varphi : S \longrightarrow \mathbb{P}^N$; si $\varphi(S)$ était une surface, on aurait $M^2 > 0$, donc $\varphi(S)$ est une courbe .

Remarques 23

1/ On peut améliorer notablement l'assertion c/ : si S est une surface minimale, le système $|nK|$ est sans points fixes dès que $n \geq 4$; dès que $n \geq 5$, le morphisme φ_{nK} est un isomorphisme en dehors de certaines courbes rationnelles, qui sont contractées en des points singuliers d'un type très simple ("singularités rationnelles"). On renvoie à [B] pour une étude très complète de la situation.

2/ Malgré le peu qu'on en a dit, les surfaces avec $\kappa = 2$ (appelées aussi "surfaces de type général" sont celles que l'on rencontre le plus fréquemment; il suffit pour s'en convaincre de regarder quelques exemples :

- Toutes les surfaces intersections complètes, sauf les V_2, V_3, V_4, $V_{2,2}$, $V_{2,3}$, $V_{2,2,2}$ sont de type général (exemple 16.2) ;

- Tout produit de courbes de genre ≥ 2 (plus généralement, toute surface fibrée sur une courbe de genre ≥ 2 , avec fibre générique de genre ≥ 2) est de type général (exemple 16.1);

- Toute surface contenue dans une variété abélienne est de type général;
- Si $f : S' \to S$ est surjectif, et si S est de type général, alors S' est de type général.

Passons maintenant aux surfaces avec $\kappa = 1$:

Théorème 24

Soit S une surface minimale avec $\kappa = 1$. Alors $K^2 = 0$, et il existe un morphisme surjectif $p : S \to B$, où B est une courbe lisse, la fibre générique de p étant une courbe elliptique lisse.

Inversement, soient S une surface minimale, B une courbe lisse, $p : S \to B$ un morphisme surjectif dont la fibre générique est elliptique. Alors l'une des 3 possibilités suivantes est réalisée :

(i) S <u>est une surface réglée de base elliptique</u>.

(ii) S <u>est une surface avec</u> $\kappa = 0$.

(iii) <u>On a</u> $\kappa = 1$; <u>alors dans</u> $\mathrm{Pic}(S) \otimes_{\mathbb{Z}} \mathbb{Q}$:

$$K \equiv \sum r_i F_{b_i} \quad , \quad r_i \in \mathbb{Q} \, , \quad r_i > 0 \quad \text{avec} \quad F_{b_i} = p^*(b_i) \, .$$

<u>Démonstration</u> : Soit S une surface minimale avec $\kappa = 1$; il résulte du lemme 8 et de la proposition 21 que $K^2 = 0$. Soit n tel que $P_n \geq 2$; notons Z la partie fixe du système $|nK|$, M sa partie mobile. Le lemme 22 montre que M définit un morphisme $q : S \longrightarrow C$. D'après le théorème de Bertini (p. 5) , il existe une factorisation :

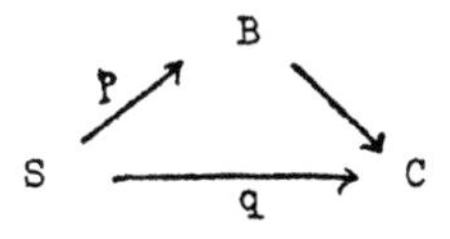

où B est une courbe lisse, et où la fibre générique F de p est lisse et irréductible. Comme F est contenue dans un diviseur de $|M|$ et que $K.M = 0$, on déduit du lemme clé que $K.F = 0$, d'où $g(F) = 1$ puisque $F^2 = 0$.

Inversement, partons d'une fibration $p : S \longrightarrow B$ dont la fibre générique F est une courbe elliptique. Soit $D = \sum n_i C_i \in |nK|$ $(n \in \mathbb{Z},\ n_i \geq 1)$ Comme $(D.F) = n(K.F) = 0$, les C_i doivent être contenus dans les fibres de p; on a donc $D^2 \leq 0$, l'égalité n'étant réalisée que si

$$D = \sum r_i F_{b_i} \, , \text{ avec } r_i \in \mathbb{Q}_+ \quad \text{(lemme 19)} .$$

Examinons maintenant la place de S dans la classification. Si S est réglée de base C , la fibre F s'envoie surjectivement sur C , donc C est elliptique ou rationnelle. Si S était rationnelle on aurait $K^2 = 8$ ou 9 (remarque 7), d'où $|-K| \neq \emptyset$ par Riemann-Roch; mais alors on trouverait $(-K)^2 \leq 0$ d'après ce qui précède, ce qui est impossible.

Si S n'est pas réglée, on a nécessairement $K^2 = 0$, donc $\kappa \leq 2$.

La dernière assertion résulte de ce qui précède.

Corollaire 25

Soit S une surface minimale avec $\kappa = 1$. Il existe un entier $d \geq 1$ tel que le système $|dK|$ soit sans points fixes.

Démonstration : Soient n tel que $P_n \neq 0$, et $D \in |nK|$; on a vu que

$D = \sum r_i F_{b_i}$, avec $r_i \in \mathbb{Q}^+$. Soit e un entier ≥ 1 tel que $er_i = m_i$ soit entier pour tout i ; alors :

$$enK \equiv \sum m_i F_{b_i} = p^* (\sum m_i b_i)$$

Pour e assez grand, le système $|\sum m_i b_i|$ sur B est sans points fixes. Il en va donc de même pour le système $|dK|$, avec $d = en$.

§ 8. Démonstration du lemme-clé

Soit S une surface minimale, C une courbe sur S telle que $(C.K) < 0$. La formule du genre montre que $C^2 \geq 0$.

a/ On a $P_n(S) = 0$ pour tout $n \geq 0$; car si $D \in |nK|$, on aurait $D.C \geq 0$ par la remarque utile (p. 6), d'où $K.C \geq 0$.

b/ Si $q \geq 1$ l'image de S dans sa variété d'Albanese est une courbe : car si C' était une surface S', on pourrait trouver une 2-forme holomorphe sur $\mathrm{Alb}(S)$ ayant une restriction à S' non identiquement nulle; par image réciproque, on en déduirait une 2-forme holomorphe non nulle sur S.

Si $q \geq 1$, il existe donc une fibration $p : S \longrightarrow B$, où B est une courbe lisse de genre q.

Premier cas : $K^2 \leq 0$.

On va supposer que S n'est pas réglée, et arriver à une contradiction.

c/ Il existe une courbe C telle que $C.K < 0$, $|C+K| = \emptyset$.

En effet le produit $(C+nK).C$ devient négatif pour n assez grand, donc par la remarque utile il existe un n tel que :

$$|C + nK| \neq \emptyset \qquad |C + (n+1)K| = \emptyset$$

Si $D = \sum n_i C_i \in |C + nK|$, on a $D.K < 0$ et $|D+K| = \emptyset$; donc il existe une courbe C_i vérifiant $C_i.K < 0$, et on a $|C_i + K| = \emptyset$.

d/ Soit C une courbe vérifiant $C.K < 0$, $|C+K| = \emptyset$. Alors $g(C) = q$; si $q \geq 2$, C est une section de la fibration $p : S \to B$; si $q = 1$, C est un revêtement étale de B .

Appliquant Riemann-Roch, on trouve en effet :

$0 = h^0(C+K) \geq 1-q + 1/2\ (C^2+C.K) = g(C) - q$

c'est-à-dire $g(C) \leq q$

Si $q = 0$, on trouve $g(C) = 0$. Si $q \geq 1$, C ne peut être contenue dans une fibre F_b de p , sans quoi on aurait $F_b = nC$ (puisque $C^2 \geq 0$: lemme 19) , d'où $g(F_b) \leq 0$ par la formule du genre, et S serait réglée. Donc C est un revêtement ramifié de B , de degré d , avec r points de ramification. La formule de Riemann-Hurwitz :

$$2g(C) - 2 = d(2q-2) + r$$

montre alors que $r = 0$, et $d = 1$ sauf si $q = 1$.

e/ Il existe une courbe C sur S vérifiant $C.K < -1$, $|C+K| = \emptyset$.

C'est clair si $q = 0$, car pour toute courbe C telle que $C.K < 0$, $|C+K| = \emptyset$, on a $g(C) = 0$ par d/ , donc $C.K \neq -1$ sans quoi C serait exceptionnelle. Supposons donc $q \geq 1$.

Soit C telle que $(C.K) < 0$, $|C+K| = \emptyset$. On voit comme en c/ qu'il existe un n tel que :

$$|2C + nK| \text{ contient un diviseur effectif } D \text{ ; } |D+K| = \emptyset .$$

On a $D.K < -1$. Posons $D = \sum_{i=1}^{r} n_i C_i$; on peut supposer que tous les C_i vérifient $C_i.K < 0$. On a aussi $|C_i + K| = \emptyset$, et par suite $g(C_i) = q$ (par d)) .

Supposons qu'il existe un i tel que $n_i \geq 2$; alors $|K+2C_i| = \emptyset$, d'où par Riemann-Roch :

$$0 = h^0(K+2C_i) \geqslant 1-q + 1/2\ (4C_i^2 + 2C_i.K) = 3(q-1) - (C_i.K)$$

ce qui est impossible.

Supposons que $r \geqslant 2$; alors $|K+C_1+C_2| = \emptyset$, d'où :

$$0 = h^0(K+C_1+C_2) = h^1(K+C_1+C_2) + 1-q + 1/2(C_1+C_2)^2 + 1/2(C_1+C_2).K$$

$$= h^1(K+C_1+C_2) + q-1 + (C_1.C_2) .$$

Ceci entraine $C_1 \cap C_2 = \emptyset$ et $h^1(K+C_1+C_2) = 0$; mais la suite exacte :

$$0 \longrightarrow \mathcal{O}_S(-C_1-C_2) \longrightarrow \mathcal{O}_S \longrightarrow \mathcal{O}_{C_1} \oplus \mathcal{O}_{C_2} \longrightarrow 0$$

donne $h^1(-C_1-C_2) = h^1(K+C_1+C_2) \geqslant 1$, d'où contradiction.

Ainsi le diviseur D est une courbe irréductible C , qui vérifie $(C.K) < -1$ et $|C+K| = \emptyset$.

f/ <u>Où l'on arrive à une contradiction</u>

Soit C une courbe vérifiant $(C.K) < -1$ et $|C+K| = \emptyset$. On a :

$$h^0(C) \geqslant 1-q + 1/2\ (C^2 - C.K) = -(C.K) \quad , \text{ i.e. } h^0(C) \geqslant 2 \ .$$

Si $q = 0$, on a donc sur S un pinceau de courbes rationnelles : on conclut par le théorème de Noether-Enriques que S est rationnelle.

Dans le cas $q \geqslant 1$, supposons d'abord que C soit une section de la fibration d'Albanese. Sa trace sur une fibre générique F est donc un point, et ce point doit bouger linéairement sur F : ceci entraine que F est rationnelle, donc que S est réglée.

Reste enfin le cas où $q = 1$ et C est un revêtement étale de B ; dans ce cas on considère la surface $S' = S \times_B C$ et la fibration $p' : S' \rightarrow C$ déduite de p . Elle possède une section naturelle $e : C \rightarrow S'$; la projection $\pi : S' \rightarrow S$ est étale, de sorte que :

$$(e(C).K_{S'}) = \deg_C(e^* K_{S'}) = \deg_C(K_S)\,|_C = (C.K) < -1 \quad \text{et}$$

$$\chi(\mathcal{O}_{S'}) = \deg(\pi)\ .\ \chi(\mathcal{O}_S) = 0$$

Le théorème de Riemann-Roch donne alors comme précédemment $h^0(S', e(C)) \geqslant 2$, de sorte que S' est réglée et donc aussi S .

<u>Deuxième cas : $K^2 > 0$</u>

g/ Le lemme 10 montre alors que $q = 0$. Comme $P_2 = 0$, le théorème de Riemann-Roch donne

$$h^o(-K) \geqslant 1+K^2 \geqslant 2$$

Supposons que $|-K|$ contienne un diviseur réductible R . Comme $R.K<0$, on peut écrire $R = C+E$, où C est une courbe irréductible telle que $(C.K)<0$ et E un diviseur effectif; mais alors $|C+K| = |-E| = \emptyset$ et on conclut comme en f/ que S est rationnelle.

On peut donc supposer désormais que tout diviseur de $|-K|$ est irréductible. Comme $(-K)^2 > 0$, la remarque utile montre que pour tout diviseur D effectif sur S il existe un n tel que :

$$|D + nK| \neq \emptyset \qquad |D + (n+1)K| = \emptyset$$

Soit $E \in |D + nK|$, et supposons d'abord $E \neq 0$. Soit C une courbe irréductible contenue dans E ; on a $|C+K| = \emptyset$, et $C.K<0$ puisque $|-K|$ contient au moins un pinceau de courbes irréductibles. On en déduit comme précédemment que S est rationnelle.

Reste le cas où quel que soit le diviseur D sur S , il existe un n tel que $D + nK \equiv 0$; autrement dit, où $\mathrm{Pic}(S) = \mathbb{Z}.[K]$. Comme $p_g = 0$, la suite exacte((a) p. 3) montre qu'on a alors $H^2(S,\mathbb{Z}) = \mathbb{Z}.[K]$; la dualité de Poincaré entraîne alors $K^2 = 1$. Mais on a :

$$\chi_{top}(S) = 2 - 2b_1 + b_2 = 3$$

$$\chi(\mathcal{O}_S) = 1$$

et il y a donc incompatibilité avec la formule de Noether.

DEUXIEME PARTIE .

§1. Structures de Hodge .

Pour tout $\mathbf{Z}$-module Γ, on notera $\Gamma_{\mathbf{R}} = \Gamma \otimes_{\mathbf{Z}} \mathbf{R}$ et $\Gamma_{\mathbf{C}} = \Gamma \otimes_{\mathbf{Z}} \mathbf{C}$.
L'espace vectoriel complexe $\Gamma_{\mathbf{C}}$ est le complexifié de $\Gamma_{\mathbf{R}}$; il est donc muni d'une conjugaison canonique, notée $x \mapsto \bar{x}$.

Définition 1 : *Une structure de Hodge de poids* n *sur un* $\mathbf{Z}$-*module de type fini* Γ *est une décomposition* :

$$\Gamma_{\mathbf{C}} = \bigoplus_{p+q=n} H^{p,q} \qquad (p,q \geq 0)$$

où les $H^{p,q}$ *sont des sous-espaces vectoriels complexes de* $\Gamma_{\mathbf{C}}$, *vérifiant* $H^{q,p} = \bar{H}^{p,q}$.

Soit X une variété projective et lisse sur $\mathbf{C}$; pour tout n, la théorie de Hodge définit une structure de Hodge de poids n sur $H^n(X,\mathbf{Z})$. Le sous-espace $H^{p,q}$ de $H^n(X,\mathbf{C})$ est alors canoniquement isomorphe à $H^q(X,\Omega^p)$; en particulier on identifie $H^o(X,\Omega^n_X)$ à un sous-espace de $H^n(X,\mathbf{C})$ (égal à $H^{n,o}$) via la cohomologie de de Rham : les n-formes holomorphes sont fermées, donc définissent des classes de cohomologie dans $H^n(X,\mathbf{C})$.

Exemple 1 : Structure de Hodge de poids 1.

Rappelons que si V est un espace vectoriel réel, il est équivalent de se donner une structure complexe sur V (i.e. un endomorphisme J de V tel que $J^2 = -1_V$) ou une décomposition de $V \otimes_{\mathbf{R}} \mathbf{C}$ en deux sous-espaces complexes conjugués : si V est muni d'une structure complexe, on décompose $V \otimes_{\mathbf{R}} \mathbf{C}$ suivant les 2 sous-espaces propres de J relatifs aux valeurs propres $+i$ et $-i$; inversement, si $V \otimes_{\mathbb{R}} \mathbb{C} = W \oplus \bar{W}$, on définit J sur V par la formule $J(w+\bar{w}) = i(w-\bar{w})$ pour tout $w \in W$.

Il en résulte que la donnée d'une structure de Hodge de poids 1 sur Γ est équivalente à la donnée d'une structure complexe sur $\Gamma_{\mathbb{R}}$, ou encore sur le tore réel $\Gamma_{\mathbb{R}}/\Gamma$. Si $\Gamma = H^1(X,\mathbb{Z})$, le tore complexe ainsi obtenu est la variété de Picard associée à X.

Exemple 2 : Structure de Hodge de poids 2 .

C'est une décomposition $\Gamma_{\mathbb{C}} = H^{2,o} \oplus H^{1,1} \oplus H^{o,2}$, avec $H^{2,o} = \bar{H}^{o,2}$ et $H^{1,1} = \bar{H}^{1,1}$. On ne sait pas associer à cette décomposition un objet géométrique aussi simple qu'un tore complexe.

Définition 2 : On appellera polarisation d'une structure de Hodge de poids n $(\Gamma, H^{p,q})$ la donnée d'une forme bilinéaire non dégénérée sur $\Gamma_{\mathbb{C}}$, symétrique si n est pair, alternée si n est impair, prenant des valeurs entières sur Γ, et vérifiant :

$(\alpha.\beta) = 0$ si $\alpha \in H^{p,q}$, $\beta \in H^{p',q'}$ et $p+p' \neq n$;

$(\alpha.\bar{\alpha}) > 0$ si $\alpha \in H^{n,o}$, n pair ;

$i(\alpha.\bar{\alpha}) > 0$ si $\alpha \in H^{n,o}$, n impair .

Si X est une variété de dimension n, le cup-produit définit une polarisation sur la structure de Hodge de $H^n(X,\mathbb{Z})$.

Exemple 1 : (suite).

Une polarisation correspond à une forme alternée sur Γ, qui étendue à $\Gamma_{\mathbb{R}}$

vérifie :

$$(J\alpha.J\beta) = (\alpha.\beta) \quad \text{et} \quad (J\alpha.\alpha) > 0 \quad \text{pour} \quad \alpha,\beta \in \Gamma_{\mathbb{R}} .$$

C'est ce qu'on appelle une forme de Riemann (ou une polarisation) sur le tore complexe $T = \Gamma_{\mathbb{R}}/\Gamma$. A partir d'une telle polarisation, la théorie des fonctions thêta fournit un plongement projectif de T, qui est donc une variété abélienne .

Exemple 2 : (Suite)

Soit $(\Gamma,H^{p,q})$ une structure de Hodge polarisée de poids 2, de sorte que $\Gamma_{\mathbb{C}} = H^{2,0} \oplus H^{1,1} \oplus H^{0,2}$. Les sous-espaces $H^{2,0}$ et $H^{0,2}$ sont isotropes; comme $(\alpha.\bar{\alpha}) > 0$ pour tout $\alpha \in H^{2,0}$, la restriction de la forme bilinéaire à $H^{2,0} \oplus H^{0,2}$ est non dégénérée, de sorte que $H^{1,1}$ est l'orthogonal de $H^{2,0} \oplus H^{0,2}$ dans $\Gamma_{\mathbb{C}}$. Il en résulte que la donnée sur Γ, muni d'un produit scalaire, d'une structure de Hodge polarisée (par le produit scalaire donné) équivaut à celle du sous-espace isotrope $H^{2,0}$ dans $\Gamma_{\mathbb{C}}$, vérifiant $(\alpha.\bar{\alpha}) > 0$ pour tout $\alpha \in H^{2,0}$.

Notons $G(\Gamma_{\mathbb{C}})$ la grassmannienne des sous-espaces isotropes de dimension $h^{2,0}$ de $\Gamma_{\mathbb{C}}$; c'est une variété algébrique projective. La condition $(\alpha.\bar{\alpha}) > 0$ pour tout $\alpha \in H^{2,0}$ définit un ouvert (analytique) $G^{o}(\Gamma_{\mathbb{C}})$ de $G(\Gamma_{\mathbb{C}})$, dont les points correspondent bijectivement aux structures de Hodge polarisées sur Γ (muni d'un produit scalaire fixé).

Soit $X \to T$ un morphisme lisse de variétés algébriques, dont les fibres sont des surfaces, et soit 0 un point de T. Posons $\Gamma = H^2(X_o,Z)$, muni de la forme d'intersection. Supposons que l'action de $\pi_1(T,0)$ sur Γ soit triviale; il existe alors pour tout $t \in T$ un isomorphisme canonique $\Gamma \to H^2(X_t,Z)$, respectant les formes d'intersection. En transportant par cet isomorphisme la structure de Hodge polarisée de $H^2(X_t,\mathbb{Z})$, on obtient pour tout t une structure de Hodge polarisée de poids 2 sur Γ, d'où une application $T \to G^o(\Gamma_{\mathbb{C}})$. La théorie de Hodge permet de prouver que cette application est analytique; on dit que c'est l'"application des périodes" de la famille de surfaces $X \to T$.

Cette terminologie est justifiée par la raison suivante. La structure de Hodge d'une surface X est déterminée par la position de $H^{2,o}$ dans $H^2(X,\mathbb{C})$; or si l'on identifie $H^{2,o}$ à $H^o(X,\Omega^2)$ et $H^2(X,\mathbb{C})$ au dual de $H_2(X,\mathbb{C})$, la flèche $i : H^o(X,\Omega^2) \to H_2(X,\mathbb{C})^*$ correspondant à l'inclusion est donnée par $\langle i(\omega),\gamma\rangle = \int_\gamma \omega$ pour $\omega \in H^o(X,\Omega^2)$, $\gamma \in H_2(X,\mathbb{Z})$. Si l'on fixe une base (ω_α) de $H^o(X,\Omega^2)$ et une base (γ_i) de $H_2(X,\mathbb{Z})$, la structure de Hodge est déterminée par la "matrice des périodes" $(\int_{\gamma_i} \omega_\alpha)$.

Si par exemple $p_g(X) = 1$, la variété $G(\Gamma_{\mathbb{C}})$ est la quadrique dans $\mathbb{P}(H^2(X,\mathbb{C}))$ définie par l'annulation de la forme quadratique sur $H^2(X,\mathbb{C})$; si l'on choisit une base (γ_i) de $H_2(X,\mathbb{Z})$, le point de cette quadrique correspondant à la structure de Hodge de $H^2(X,\mathbb{Z})$ a pour coordonnées $\int_{\gamma_i} \omega$, où ω est une 2-forme $\neq 0$ sur X.

Problème de Torelli .

R. Torelli a démontré ([T]) que la donnée des périodes d'une surface de Riemann X (i.e. de la structure de Hodge polarisée sur $H^1(X,\mathbb{Z})$, ou encore de la jacobienne polarisée JX) caractérise X à isomorphisme près. On appelle maintenant "problème de Torelli" la question de savoir si les structures de Hodge polarisées sur la cohomologie d'une variété déterminent la variété. La solution de ce problème très difficile n'est connue que dans de rares cas.

Le cas des surfaces K3 a été résolu récemment par Chafarevitch et Piatechki-Chapiro ([Ch-P]); on va donner ici une indication de leur démonstration. On renvoie pour les détails (et des compléments) à l'article [Ch-P].

Etudions d'abord le problème de Torelli pour les surfaces abéliennes. La structure de Hodge de poids un d'une telle surface A détermine sa variété de Picard, c'est à dire la variété abélienne duale $\hat{A}$; il est bien connu que l'on récupère A comme variété abélienne duale de $\hat{A}$. On a également un théorème de Torelli pour les structures de Hodge de poids 2 :

Proposition 3 :

Soient A, A' deux surfaces abéliennes. Si les structures de Hodge polarisées sur $H^2(A,\mathbb{Z})$ et $H^2(A',\mathbb{Z})$ sont isomorphes, A et A' sont isomorphes.

Démonstration : Rappelons qu'il existe un isomorphisme canonique de $\Lambda^2 H^1(A,\mathbb{Z})$ sur $H^2(A,\mathbb{Z})$; avec cette identification, la décomposition de Hodge de $H^2(A,\mathbb{C})$ est donnée par :

$$H^2(A,\mathbb{C}) \cong \Lambda^2 H^1(A,\mathbb{C}) = \Lambda^2 H^{1,0} \oplus (H^{1,0} \otimes H^{0,1}) \oplus \Lambda^2 H^{0,1} .$$

Il faut donc montrer que la position de $H^{1,0}$ dans $H^1(A,\mathbb{C})$ est caractérisée par celle de $\Lambda^2 H^{1,0}$ dans $\Lambda^2 H^1(A,\mathbb{C})$; or, c'est là un résultat bien connu: il exprime que le "morphisme de Plücker", qui va de la grassmannienne des droites de $\mathbb{P}^3$ dans $\mathbb{P}^5$, est un plongement.

§2. Surfaces de Kummer et surfaces K3 .

Rappelons (Théorème 18, 1ère partie) qu'une surface K3 est une surface vérifiant $K \equiv 0$, $q = 0$. Les intersections complètes V_4, $V_{2,3}$, $V_{2,2,2}$ sont des surfaces K3 (cf. exemple 16, 1ère partie); plus généralement, pour tout $g \geq 3$ il existe une famille de surfaces K3 de degré $2g-2$ dans P^g. Nous allons indiquer une construction particulière de surfaces K3, à partir de surfaces abéliennes.

Soit A une surface abélienne. L'involution θ de A, définie par $\theta(a) = -a$, admet 16 points fixes isolés $a_1,\ldots,a_{16}$, qui sont les points d'ordre 2 de A. Notons $\varepsilon : \hat{A} \to A$ l'éclatement de ces 16 points, E_i le diviseur exceptionnel $\varepsilon^{-1}(a_i)$ $(1 \leq i \leq 16)$. L'involution θ se prolonge en une involution σ de $\hat{A}$. On désigne par X la variété quotient $\hat{A}/\sigma$, par $\pi : \hat{A} \to X$ l'application canonique; on pose $L_i = \pi(E_i)$.

Proposition 4. La surface $X = \hat{A}/\sigma$ est une surface K3, appelée surface de Kummer associée à A.

Démonstration : Montrons d'abord que X est lisse. C'est clair en dehors des L_i; soient $p \in E_i$, $q = \pi(p)$. On peut trouver des coordonnées locales (x,y) sur A au voisinage de a_i telles que :

$$\theta^* x = -x \qquad \theta^* y = -y$$

et que x et $t = \frac{y}{x}$ forment un système de coordonnées locales dans $\hat{A}$ au voisinage de p. Comme $\sigma^* t = t$, on conclut que t et $u = x^2$ forment un système de coordonnées locales sur X au voisinage de q; en particulier X est lisse en q.

Soit ω une 2-forme holomorphe non nulle sur A. La forme $\varepsilon^*\omega$ est invariante par σ; elle est donc de la forme $\pi^*\alpha$, où α est une 2-forme méromorphe sur X. Il est clair que α est holomorphe et partout $\neq 0$ en dehors des L_i; au voisinage de a_i, on a :

$$\omega = k\, dx \wedge dy \qquad k \in \mathbb{C}$$

d'où, au voisinage de p : $\varepsilon^*\omega = k\, dx \wedge d(tx) = kx.dx \wedge dt = \frac{k}{2}\,\pi^*(du \wedge dt)$, ce qui montre que $\alpha = \frac{k}{2}.du \wedge dt$ est holomorphe $\neq 0$ sur L_i. Ainsi α est une 2-forme holomorphe partout $\neq 0$ sur X, donc $K_X \equiv 0$. S'il existait une 1-forme holomorphe non nulle sur X, on en déduirait une 1-forme holomorphe $\neq 0$ sur A invariante par θ, ce qui est impossible : donc $q(X) = 0$ et X est une surface K3 .

La surface de Kummer X contient 16 droites rationnelles $L_1,\ldots,L_{16}$, vérifiant $L_i.L_j = 0$ pour $i\neq j$. Inversement :

Lemme 5. *Soient X une surface K3, $L_1,\ldots,L_{16}$ 16 courbes lisses rationnelles telles que $L_i \cap L_j = \emptyset$ pour $i\neq j$. Supposons qu'il existe $l \in \mathrm{Pic}(X)$ tel que $2l = \Sigma L_i$ dans Pic(X). Alors X est une surface de Kummer.*

Démonstration : Notons L le fibré en droites de classe l; soit s une section du fibré $L^{\otimes 2}$ dont le diviseur des zéros est ΣL_i.

Posons : $\hat{A} = \{x \in L,\ x^{\otimes 2} = s\}$.

La projection de L sur X définit un revêtement double $\pi : \hat{A} \to X$, ramifié le long des L_i. On a $\pi^*L_i = 2E_i$, où E_i est une courbe rationnelle lisse; de plus :

$$4E_i^2 = (\pi^*L_i)^2 = 2\,L_i^2 = -4 \qquad \text{d'où } E_i^2 = -1.$$

Le critère de contraction de Castelnuovo montre qu'il existe un morphisme birationnel $\varepsilon : \hat{A} \to A$ qui contracte les E_i . Prouvons que A est une surface abélienne. On a :

$$K_{\hat{A}} \equiv \varepsilon^* K_A + \Sigma E_i \equiv \pi^* K_X + \Sigma E_i \ , \qquad \text{d'où } K_A \equiv 0 \ .$$

Calculons $\chi(\mathcal{O}_A) = \chi(\mathcal{O}_{\hat{A}})$. On vérifie facilement que $\pi_* \mathcal{O}_{\hat{A}} = \mathcal{O}_X \oplus L^{-1}$; par suite $\chi(\mathcal{O}_A) = \chi(\mathcal{O}_X) + \chi(L^{-1}) = 4 + \frac{1}{2} l^2$ par Riemann-Roch.

Or $l^2 = \frac{1}{4} (\Sigma L_i)^2 = - 8$; on a donc $\chi(\mathcal{O}_A) = 0$, d'où $q(A) = 2$.

La classification des surfaces (Théorème 18, 1ère partie) montre que A est une surface abélienne.

Notons σ l'involution qui échange les deux feuillets du revêtement ramifié π; comme $\sigma(E_i) = E_i$, elle provient d'une involution θ de A. Soit $a_i = \varepsilon(E_i)$ l'un des points fixes de θ; θ opère sur l'espace tangent à A en a_i par multiplication par -1. On en déduit aussitôt que θ est l'involution $a \mapsto -a$ de A, pour la structure de groupe de A pour laquelle a_i est l'origine. Par suite, X est la surface de Kummer associée à A.

<u>Lemme 6</u>. <u>Avec les notations précédentes, désignons par</u> L_X <u>le sous-groupe de</u> $H^2(X,\mathbb{Z})$ <u>engendré par les</u> L_i. <u>Le sous-groupe</u> $\pi_* \varepsilon^* H^2(A,\mathbb{Z}) \subset H^2(X,\mathbb{Z})$ <u>est l'orthogonal de</u> L_X <u>dans</u> $H^2(X,\mathbb{Z})$.

<u>Démonstration</u> : Au cours de la démonstration, on notera pour abréger $H^i(T) = H^i(T,\mathbb{Z})$ pour toute variété T.

Il est immédiat que $\pi_* \varepsilon^* H^2(A) \subset (L_X)^{\perp}$; montrons l'inclusion contraire. Soit $x \in (L_X)^{\perp}$; comme $\pi^* x$ est orthogonal aux E_i, il existe $a \in H^2(A)$ tel que $\pi^* x = \varepsilon^* a$. Comme $x = \frac{1}{2} \pi_* \pi^* x$, il suffit de prouver que $a = 2a'$ pour un $a' \in H^2(A)$; ou encore, par dualité de Poincaré, que $(a.b)$ est pair pour tout $b \in H^2(A)$.

Posons $A' = \hat{A} - \cup E_i$, $X' = \hat{X} - \cup L_i$; notons $\pi' : A' \to X'$ la restriction de π. Considérons le diagramme de suites exactes :

$$\begin{array}{ccccccccc}
0 & \longrightarrow & \mathbf{Z}^{16} & \xrightarrow{(L_i)} & H^2(X) & \longrightarrow & H^2(X') & \longrightarrow & 0 \\
 & & \downarrow{\scriptstyle 2} & & \downarrow{\scriptstyle \pi^*} & & \downarrow{\scriptstyle \pi'^*} & & \\
0 & \longrightarrow & \mathbf{Z}^{16} & \xrightarrow{(E_i)} & H^2(\hat{A}) & \longrightarrow & H^2(A') & \longrightarrow & 0
\end{array}$$

On va montrer plus bas que π'^* est surjectif; on en déduit que $H^2(\hat{A})$ est engendré par $\mathrm{Im}(\pi^*)$ et les E_i. Posons $\varepsilon^* b = \pi^* y + \Sigma n_i E_i$, avec $y \in H^2(X)$, $n_i \in \mathbf{Z}$; on a :

$$(a.b) = (\varepsilon^* a.\varepsilon^* b) = (\varepsilon^* a.\pi^* y) = (\pi^* x.\pi^* y) = 2(x.y)$$

d'où le résultat.

Il reste à montrer que π'^* est surjectif. Puisque π' est un revêtement étale de degré 2, il existe une suite spectrale :

$$E_2^{pq} = H^p(\mathbf{Z}/(2), H^q(A')) \Longrightarrow H^{p+q}(X') .$$

Notons σ_q l'action de l'involution σ (restreinte à A') sur $H^q(A')$, et $\varepsilon_p = (-1)^p . \mathrm{Id}_{H^q(A')}$. La théorie de la cohomologie des groupes finis donne : $E_2^{pq} = \mathrm{Ker}(\sigma_q - \varepsilon_p) / \mathrm{Im}(\sigma_q + \varepsilon_p)$ pour $p \neq 0$.

Pour $q \leq 2$, la restriction $H^q(A) \to H^q(A')$ est un isomorphisme; par suite $\sigma_q = \varepsilon_q$ pour $q \leq 2$. On en déduit en particulier $E_2^{2,1} = E_2^{3,0} = 0$; il en résulte que l'"edge-homomorphisme" $H^2(X') \to E_2^{0,2}$ de la suite spectrale est surjectif. Mais cet homomorphisme n'est autre que π'^*; ceci achève de démontrer le lemme .

Proposition 7. Soient X et X' deux surfaces de Kummer, φ : $H^2(X,\mathbf{Z}) \to H^2(X',\mathbf{Z})$ un isomorphisme de structures de Hodge polarisées, tel que

$\varphi(L_X) = L_{X'}$. Alors X et X' sont isomorphes.

Démonstration : Notons A,A' les surfaces abéliennes associées à X,X'. Il est immédiat que $(\pi_*\varepsilon^*a.\pi_*\varepsilon^*b) = 2(a.b)$ pour $a,b \in H^2(A,Z)$; en particulier $\pi_*\varepsilon^*$ est injectif. Il résulte alors du lemme 6 que φ induit un isomorphisme ψ de $H^2(A,Z)$ sur $H^2(A',Z)$, respectant le cup-produit. Il est clair que ψ induit un isomorphisme de structures de Hodge; la proposition 3 montre alors que A et A' sont isomorphes. Elles sont par conséquent isomorphes comme variétés abéliennes, ce qui entraîne que X et X' sont isomorphes.

§3. Surfaces de Kummer spéciales.

On dit qu'une surface de Kummer associée à une surface abélienne A est spéciale si A est réductible, i.e. s'il existe une suite exacte de variétés abéliennes :

$$0 \longrightarrow E' \longrightarrow A \xrightarrow{p} E \longrightarrow 0$$

où E et E' sont des courbes elliptiques.

Notons R la courbe (rationnelle) quotient de E par l'involution $x \longmapsto -x$; on a un diagramme commutatif :

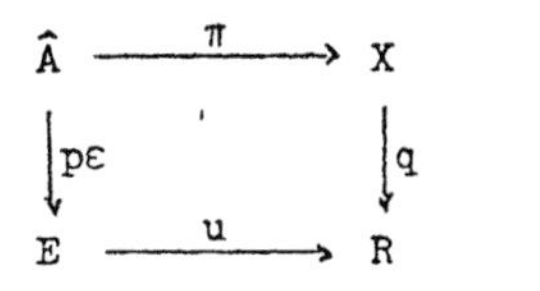

On dira que la fibration $q : X \to R$ est une "fibration de Kummer" pour X. Il est clair que q est lisse en dehors des 4 points de ramification $r_1,\dots,r_4$ de u.

Proposition 8.

Soient e *un point d'ordre* 2 *de* E, $a_1,\ldots,a_4$ *les* 4 *points d'ordre* 2 *de* A *tels que* $p(a_i) = e$; *on pose* $r = u(e)$. *On a* :

$$q^*r = 2\, F_r + L_1 + L_2 + L_3 + L_4$$

où F_r *est une courbe rationnelle lisse et* $F_r.L_i = 1$ $(1 \leq i \leq 4)$.

Démonstration : On a :

$$q^*r = \frac{1}{2}\pi_*\pi^*q^*r = \pi_*\varepsilon^*p^*e = \pi_*\ (\hat{F}_e + \sum_{i=1}^{4} E_i)\ ,$$

où $\hat{F}_e$ est le transformé strict dans $\hat{A}$ de la courbe elliptique $F_e = p^{-1}(e)$. L'involution σ induit sur $\hat{F}_e$ une involution du type $x \longmapsto -x$; par suite, $\pi_*\hat{F}_e = 2F_r$, où $F_r = \hat{F}_e/(\sigma)$ est isomorphe à P^1.

Corollaire 9.

Soient X (*resp.* X') *une surface de Kummer spéciale*, f (*resp.* f') *la fibre générique d'une fibration de Kummer. Soit* φ *un isomorphisme de* $H^2(X,Z)$ *sur* $H^2(X',Z)$, *induisant un isomorphisme des structures de Hodge polarisées, transformant cycles effectifs en cycles effectifs, et tel que* $\varphi(f) = f'$. *Alors* X *et* X' *sont isomorphes.*

Démonstration : Compte tenu de la proposition 7, il suffit de prouver que $\varphi(L_X) = L_{X'}$. On a $f = 2F_r + \sum_{i=1}^{4} L_i$ dans Pic(X), d'où $f' = 2\varphi(F_r) + \Sigma\varphi(L_i)$ dans Pic(X'). Comme les seules décompositions de f' en somme de diviseurs effectifs sont de la forme $f' = 2F_{r'} + \Sigma L'_j$, on en déduit que $\varphi(L_i) = L'_j$, d'où le résultat.

Pour $f \in \mathrm{Pic}(X)$, on notera $\tilde{S}_X(f)$ le sous-groupe de Pic(X) engendré par les élements x tels que $x^2 = -2$, $x.f = 0$; si $f \in \tilde{S}_X(f)$, on pose $S_X(f) = \tilde{S}_X(f)/\mathbb{Z}.f$, et on le munit de la forme quadratique induite. Supposons que f soit la fibre d'une fibration $q: X \to \mathbb{P}^1$, à fibres connexes; soit x un élément de Pic(X) tel que $x^2 = -2$ et $x.f = 0$. Le théorème de Riemann-Roch montre que l'une des classes x ou $-x$ contient un diviseur effectif, qui est nécessairement une somme de composantes des fibres de q. On en déduit que dans ce cas le module $\tilde{S}_X(f)$ est engendré par les composantes des fibres de q.

Notons par ailleurs D_4 le module $\mathbb{Z}^4$ muni du produit scalaire défini par:

$$e_o.e_i = 1 \quad \text{pour} \quad 1 \leq i \leq 3$$
$$e_i^2 = -2 \quad \text{pour} \quad 0 \leq i \leq 3$$
$$e_i.e_j = 0 \quad \text{pour} \quad 1 \leq i \leq j \leq 3 .$$

Il résulte aussitôt de la proposition 8 que lorsque f est la fibre d'une fibration de Kummer, $S_X(f)$ est isomorphe comme module quadratique à $(D_4)^4$.

Proposition 10.

Soient X une surface K3, $q: X \to \mathbb{P}^1$ un morphisme surjectif à fibres connexes, f la classe dans Pic(X) d'une fibre de q. Supposons que $S_X(f) \cong (D_4)^4$; alors, X est une surface de Kummer spéciale, et q est une fibration de Kummer.

Démonstration : Il résulte aussitôt du lemme 19, 1ère partie, que la forme quadratique sur $S_X(f)$ est négative non dégénérée. Le module quadratique $S_X(f)$ se décompose donc de manière unique comme somme de sous-modules irréductibles; chacun de ces sous-modules est engendré par les composantes d'une fibre réductible.

Sous l'hypothèse de l'énoncé, on conclut que q admet quatre fibres réductibles, de type D_4; c'est à dire que chaque fibre réductible s'écrit :

$$F_k = 2F'_k + \sum_{i=1}^{4} L_{4k+i} \qquad k = 0,\ldots,3 .$$

On obtient en particulier 16 courbes rationnelles lisses L_i, deux à deux sans point commun, et telles que $\Sigma L_i = -2(\Sigma F'_k) + 4f$. On déduit alors du lemme 5 que X est la surface de Kummer associée à une surface abélienne A.

Considérons le revêtement ramifié $\pi : \hat{A} \to X$, et posons $E'_o = \pi^{-1}(F'_o)$. C'est un revêtement double de $\mathbb{P}^1$ ramifié en 4 points, donc une courbe elliptique, qui coupe transversalement les droites exceptionnelles $E_1,\ldots,E_4$. Par suite, $\varepsilon(E'_o) = E'$ est une courbe elliptique sur A; en choisissant une origine convenable sur E' et A, on peut considérer E' comme une sous-variété abélienne de A. Notons E la courbe elliptique A/E'; on a un diagramme commutatif :

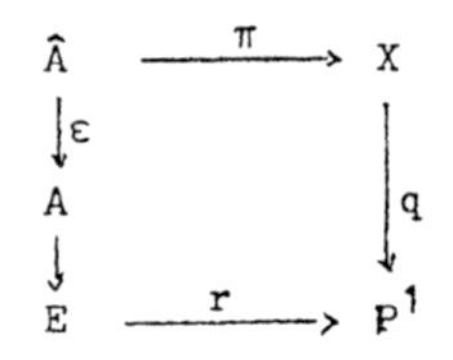

$$\begin{array}{ccc} \hat{A} & \xrightarrow{\pi} & X \\ \downarrow \varepsilon & & \\ A & & \downarrow q \\ \downarrow & & \\ E & \xrightarrow{r} & \mathbb{P}^1 \end{array}$$

où r est un revêtement double, ramifié au-dessus des 4 points de $\mathbb{P}^1$ au-dessus desquels q n'est pas lisse. On en conclut que q est une fibration de Kummer pour X.

Lemme 11 .

Soit X une surface K3, et soit $f \in \mathrm{Pic}(X)$ une classe primitive (i.e. $f=mf'$ pour m entier entraîne $m=\pm 1$), telle que $f^2=0$ et $(f.x) \geq 0$ pour tout diviseur effectif x. Alors f est la fibre d'un morphisme $q: X \to \mathbb{P}^1$ à fibres connexes.

Démonstration : Comme $(f.h) \geq 0$ si h est la classe d'une section hyperplane,

$-f$ ne peut être la classe d'un diviseur effectif. Le théorème de Riemann-Roch montre alors que $h^o(f) \geq 2$. Ecrivons $f \equiv Z+M$, où Z est fixe et M n'a pas de composantes fixes. L'argument du lemme 22, 1ère partie, montre que $Z^2=M^2=0$. On en déduit que $Z=0$ (sans quoi Z serait mobile par Riemann-Roch) et que $|M|$ est un pinceau sans points fixes. Il définit donc un morphisme $q : X \to P^1$, qui se factorise (théorème de Bertini) en: $q : X \xrightarrow{p} C \xrightarrow{r} P^1$, où p est à fibres connexes. Comme $q(X) = 0$, la courbe C est isomorphe à P^1. On a donc $f = d.f'$, où f' est la fibre de p et d le degré du morphisme r. Mais l'hypothèse entraîne $d=1$, d'où le résultat.

Théorème 12.

Soient X une surface de Kummer spéciale, X' une surface K3, φ : $H^2(X,\mathbb{Z}) \to H^2(X',\mathbb{Z})$ un isomorphisme de structures de Hodge polarisées, transformant cycles effectifs en cycles effectifs. Alors X et X' sont isomorphes.

Démonstration : Notons f la classe dans $\mathrm{Pic}(X)$ de la fibre d'une fibration de Kummer; considérons l'élément $f' = \varphi(f) \in \mathrm{Pic}(X')$. Il vérifie les hypothèses du lemme 11; il définit donc une fibration $q: X \to P^1$, à fibres connexes, et on a $S_{X'}(f') \cong (D_4)^4$. Il résulte alors de la proposition 10 que X' est une surface de Kummer spéciale et q une fibration de Kummer. On conclut avec le corollaire 9.

Remarque 13. Soient X, X' deux surfaces K3, φ: $\mathrm{Pic}(X) \to \mathrm{Pic}(X')$ un isomorphisme respectant le cup-produit, h la classe dans $\mathrm{Pic}(X)$ d'une section hyperplane de X. Les conditions suivantes sont équivalentes :

a) φ transforme les cycles effectifs en cycles effectifs;

b) $\varphi(h)$ est la classe d'un diviseur ample.

En effet, cela résulte facilement du fait que les classes de diviseurs effectifs sur X sont les sommes d'éléments $x \in \text{Pic}(X)$ tels que : $x^2 \geq -2$, $(x.h) \geq 0$.

Théorème 14.

Soient X (resp. X') une surface K3, h (resp. h') la classe d'une section hyperplane. On suppose qu'il existe un isomorphisme $\varphi : H^2(X,\mathbb{Z}) \to H^2(X',\mathbb{Z})$, induisant un isomorphisme des structures de Hodge polarisées, tel que $\varphi(h)=h'$. Alors, X et X' sont isomorphes.

Idée de la démonstration : On construit un espace analytique M_g, qui est un espace des modules grossier pour les surfaces K3 munies d'une "polarisation" h telle que $h^2 = 2g-2$. Pour cela, on considère le schéma de Hilbert qui paramètre les surfaces de degré $2g-2$ dans P^g qui sont des surfaces K3, et on le divise par l'action des automorphismes de $\mathbb{P}^g$.

Soit d'autre part Γ un module quadratique isomorphe au H^2 d'une surface K3, η un élément de Γ tel que $\eta^2 = 2g-2$. On désigne par $G^\eta(\Gamma_\mathbb{C})$ la sous-variété de $G^o(\Gamma_\mathbb{C})$ (§1) formée des sous-espaces $H^{2,0}$ de $\Gamma_\mathbb{C}$ qui sont orthogonaux à η. Notons D le groupe (discret) des automorphismes de Γ qui fixent η et P l'espace analytique quotient $G^\eta(\Gamma_\mathbb{C})/D$. En choisissant arbitrairement, pour toute surface K3 polarisée (X,h), un isomorphisme u de $H^2(X,\mathbb{Z})$ sur Γ tel que $u(h) = \eta$, et en transportant sur Γ la structure de Hodge de $H^2(X,\mathbb{Z})$ par cet isomorphisme, on obtient une application des périodes $p : M_g \to P$, qui est analytique.

L'énoncé du théorème signifie que p est injectif. Or, le théorème 12 (joint à la remarque 13) montre que $p^{-1}(p(x)) = \{x\}$ lorsque le point x de M_g

correspond à une surface de Kummer spéciale. On montre alors que l'ensemble des $p(x)$, où x correspond à une surface de Kummer spéciale, est dense dans P. L'injectivité de p en résulte aussitôt.

BIBLIOGRAPHIE

[B] E. BOMBIERI : Canonical models of surfaces of general type. Publ. Math. I.H.E.S. 42 (1973)

[Be] A. BEAUVILLE : Surfaces algébriques complexes. A paraître dans Astérisque.

[Ch1] I.R. CHAFAREVITCH : Foundations of algebraic geometry. Springer-Verlag

[Ch2] CHAFAREVITCH et al. : Algebraic surfaces. Proc. of Steklov inst. of Math. n°75 (1965).

[Ch-P] CHAFAREVITCH et PIATECHKI-CHAPIRO : A Torelli theorem for algebraic surfaces of type K3. Math. USSR Izvestja . Vol.5 (1971), n°3.

[G] A. GROTHENDIECK : Techniques de construction en géométrie analytique. I-X Sém. Cartan, t.13 (1960/61).

[R] M. RAYNAUD : Familles de fibrés vectoriels sur une surface de Riemann. Sém. Bourbaki n°316 (1966/67).

[T] R. TORELLI : "Sulle varietà di Jacobi" . Rendiconti della R. Acc. dei Lincei, série 5ª, Vol.22 (1913).

Arnaud Beauville

Université d'Angers

Fac. des Sciences
Bd. Lavoisier

49045 - ANGERS Cédex

CENTRO INTERNAZIONALE MATEMATICO ESTIVO

(C.I.M.E.)

METHODS OF ALGEBRAIC GEOMETRY IN CHAR. p AND THEIR APPLICATIONS

ENRICO BOMBIERI

METHODS OF ALGEBRAIC GEOMETRY IN CHAR. p AND THEIR APPLICATIONS TO ENRIQUES' CLASSIFICATION

Enrico Bombieri
Scuola Normale Superiore, Pisa

I. The aim of these lectures is to illustrate some of the algebraic techniques needed for algebraic geometry in char. p, with a particular view to the theory of algebraic surfaces and Enriques' classification. We shall study the new char. p features of Kodaira's vanishing theorem, the completeness of the characteristic system and the theory of the Picard variety, together with the study of elliptic and quasi-elliptic fibrations, the study of Enriques' surfaces in char. 2 and the characterization of abelian surfaces by means of their numerical invariants.

Our basic list of notations will be as follows:

X = an algebraic variety

$\mathrm{Alb}\, X$ = Albanese variety of X

$\mathrm{Pic}\, X$ = Picard scheme of X

$\mathrm{Pic}^0\, X$ = connected component of $0 \in \mathrm{Pic}\, X$

q = $\dim \mathrm{Pic}\, X$ = $\dim \mathrm{Alb}\, X$, the irregularity of X

$K = K_X$ = the canonical class of X

B_i = i^{th} Betti number

$h^{p,q} = \dim H^q(X, \Omega^p)$

$p_g = h^{0,2} = h^{2,0}$ if $\dim X = 2$

$\omega_X = \Omega_X^2$ if $\dim X = 2$ and X is smooth = in general, the dualizing sheaf for Gorenstein varieties

$h^q(F) = \dim H^q(X, F)$

II. The Kodaira vanishing theorem.

The origin of Kodaira's vanishing theorem goes back to the theory of Riemann surfaces: the problem there is the construction of meromorphic functions with prescribed zeros and poles. For example, it is a well-known fact in the theory of elliptic functions that a meromorphic non-constant doubly periodic function of a complex variable must have at least two poles or a double pole, that is a non-constant meromorphic function on a curve of genus 1 has at least two poles.

Let us consider in more detail the case of a projective non-singular curve C over an algebraically closed field k. Suppose $g = \text{genus}(C)$ and let $P_1,\ldots,P_r$ be r points on C, and let $m_1,\ldots,m_r$ be arbitrary integers. We ask what is the dimension of the vector space of meromorphic functions on C, regular outside $P_1,\ldots,P_r$ and having their poles of multiplicity at most $m_1,\ldots,m_r$. The Riemann-Roch theorem shows that if $m_1 + \cdots + m_r$ is sufficiently large then

$$\dim = m_1 + \cdots + m_r + 1 - g \tag{2.1}$$

and in fact sufficiently large means here $m_1 + \cdots + m_r > 2g - 2$. In old-fashioned language, one says that $\Sigma m_i + 1 - g$ is the postulated dimension of this space and eq. (2.1) is referred to as a postulation formula. The problem arises: when is the dimension of this space actually equal to what it is postulated to be?

This is better stated in terms of invertible sheaves. If D is the divisor $D = \Sigma m_i P_i$, and if we set $\deg D = \Sigma m_i$ and if L is the invertible sheaf

$L = \mathcal{O}(D)$ = sheaf of germs of functions on C with poles not worse than $-D$

$\cong$ sheaf of germs of sections of the line bundle $[D]$ determined by D

then clearly our vector space is $H^0(C,L)$ and our postulation formula becomes

$$\dim H^0(C,L) = \deg D + 1 - g$$
$$= \deg L + 1 - g .$$

As we have seen, this is not universally true; what is true is

(2.2) $$\dim H^0(C,L) - \dim H^1(C,L) = \deg L + 1 - g .$$

The question of the validity of the postulation formula now becomes that of the vanishing of the cohomology group $H^1(C,L)$. In order to handle this group, the Roch part of the Riemann-Roch theorem comes to help. The point is that the dual of the vector space $H^1(C,L)$ is isomorphic (via differentials on C) to the space $H^0(C,\omega_C \otimes L^{-1})$, so that we ask about the vanishing of this last space. Now recall that a non-zero meromorphic function on a curve C (projective!) has as many zeros as poles. If $f \in H^0(C,\omega_C \otimes L^{-1})$ we thus obtain

$$\deg \operatorname{div}(f) = 0 \quad \text{if} \quad f \neq 0,$$

but also the assumption $f \in H^0(C,\omega_C \otimes L^{-1})$ shows that

$$\deg \operatorname{div}(f) \geqq - \deg (\omega_C \otimes L^{-1})$$
$$= \deg L - (2g - 2) .$$

This is a contradiction if $\deg L > 2g - 2$, hence in this case every section of $\omega_C \otimes L^{-1}$ is 0. This shows indeed that

(2.3) $$H^1(C,L) = 0 \quad \text{if} \quad \deg L > 2g - 2;$$

the result (2.3) is a form of Kodaira's vanishing theorem for the curve C. Another form of this result can be stated as

(2.4) $$H^0(C,L^{-1}) = 0 \quad \text{if} \quad \deg L > 0$$

and it is indeed (2.4) that we proved with the previous argument.

In the case of surfaces, we deal with functions having zeros and poles along given curves. Assuming X a projective non-singular surface, L an invertible sheaf on X, we have the Riemann-Roch theorem

(2.5) $$\chi(L) = \sum_{i=0}^{2} (-1)^i \dim H^i(X,L)$$
$$= h^0(L) - h^1(L) + h^2(L)$$
$$= \frac{1}{2}(L\cdot(L - K)) + \frac{1}{12}(K_X^2 + c_2[X])$$

where $(A \cdot B)$ denotes the intersection number and where

$$c_2[X] = \text{value on } X \text{ of the second Chern class of the tangent bundle}$$
$$= \sum_{i=0}^{4} (-1)^i B_i = \text{Euler class}$$
$$= \text{self-intersection of the diagonal in } X \times X.$$

The additional basic results used together with the Riemann-Roch theorem are:

A. Serre's duality. If $n = \dim X$ then $H^i(X,L)$ is dual to $H^{n-i}(X,\omega_X \otimes L^{-1})$;

B. The theorem of Enriques-Severi-Zariski-Serre. If L is ample (i.e. if there is an integer ν such that $\nu > 0$ and $L^{\otimes\nu} \cong \mathcal{O}(D)$, D hyperplane section of X in some projective embedding) then for every coherent sheaf F we have

$$H^i(X, F \otimes L^m) = 0$$

for all sufficiently large m, i.e. $m \geqq m_0(F,L)$, and all $i > 0$.

C. Kodaira's vanishing theorem. (char = 0)

(first form) If L is ample then

$$H^i(X,\omega_X \otimes L) = 0 \quad \text{for } i > 0 .$$

(second form) If L is ample then

$$H^i(X,L^{-1}) = 0 \quad \text{for } i < \dim X .$$

More generally, we have

C'. The theorem of Kodaira-Akizuki-Nakano. (char = 0)

If L is ample then

$$H^q(X,\Omega^p \otimes L^{-1}) = 0 \quad \text{for } p + q < \dim X .$$

If we compare B and C (first form) we see that Kodaira's vanishing theorem is the sharpest form of the theorem of Enriques-Severi-Zariski-Serre in the special case in which $F = \omega_X$. On the other hand, the condition char = 0 is really required and additional conditions are needed in char $\neq 0$ for the validity of C.

Now we shall give Ramanujam's proof of the result of Kodaira-Akizuki-Nakano.

Let us consider first a <u>special case:</u> L is very ample, hence there is a section $s \in H^0(X,L)$ with $D = \operatorname{div}(s)$ = a smooth hyperplane section of X.

We choose local coordinates $x_1,\dots,x_n$ on X such that $x_1,\dots,x_{n-1}$ are local coordinates on D and $x_n \equiv f(x_1,\dots,x_{n-1},x_n) = 0$ is a local equation for D, at a point $P \in D$. We have the exact sequence of sheaves on X

$$(2.6) \qquad 0 \longrightarrow \Omega_D^{q-1} \otimes \mathcal{J}_D \otimes \mathcal{O}_D \xrightarrow{df} \Omega_X^q \otimes \mathcal{O}_D \xrightarrow{res} \Omega_D^q \longrightarrow 0$$

where $\mathcal{J}_D$ is the sheaf of ideals of D (thus $\mathcal{J}_D \cong \mathcal{O}(-D)$), $\mathcal{O}_D \cong \mathcal{O}_X/\mathcal{J}_D$, $\xrightarrow{df}$ is the multiplication by df and res is the restriction map. It is a useful exercise to verify the asserted exactness, and the fact that the mappings above are well-defined (i.e. independent of local coordinates and patching well together). We do this as follows. Since the sheaves in question are supported at D, we need verify exactness only at points $P \in D$. First of all, we check that the map df is well-defined. If $\{f_i,U_i\}$ is a set of local equations for D, $f_i = g_{ij}f_j$ on $U_i \cap U_j$, so that $\{g_{ij},U_i \cap U_j\}$ is a 1-cocycle defining D, we see that the ideal sheaf $\mathcal{J}_D$ is an invertible sheaf with transition functions $\{g_{ij}^{-1},U_i \cap U_j\}$. Now

$$df_i = g_{ij}\, df_j \mod f_j$$

hence the map

$$\Omega_D^{q-1} \otimes \mathcal{J}_D \otimes \mathcal{O}_D \xrightarrow{df} \Omega_X^q \otimes \mathcal{O}_D$$

is well-defined. To verify exactness, we have that $\Omega_{X,P}^q$ consists of germs of differential forms $\Sigma\varphi_I\, dx_I$, $I = (i_1,\dots,i_q)$, $1 \leq i_\nu \leq n$, $dx_I = dx_{i_1} \wedge \dots \wedge dx_{i_q}$, $(\Omega_X^q \otimes \mathcal{O}_D)_P$ is identified with elements $\Sigma(\varphi_I \mod (x_n))dx_I$, and the kernel of the restriction map is made up of elements $\sum\limits_{\substack{n \notin I \\ \text{card } I=q-1}} (\varphi_I \mod (x_n))dx_I \wedge dx_n$ which via $\xrightarrow{df}$ is identified with $\sum\limits_{\substack{n \notin I \\ \text{card } I=q-1}} (\varphi_I \mod (x_n))dx_I$, i.e. with $(\Omega_D^{q-1} \otimes \mathcal{J}_D \otimes \mathcal{O}_D)_P$, and the result follows.

As we are dealing with algebraically closed fields of definition, of char. 0 we shall suppose that $k = \mathbf{C}$. Now, since D is a hyperplane section of X, the Lefschetz theorem shows that the natural map $H^i(X,\mathbf{C}) \longrightarrow H^i(D,\mathbf{C})$ is an isomorphism if $i < n - 1$ and injective if $i = n - 1$. By the Hodge decomposition

$$H^i(X,\mathbf{C}) \cong \bigoplus_{p+q=i} H^p(X,\Omega_X^q)$$

one deduces that

$$H^p(X,\Omega_X^q) \longrightarrow H^p(D,\Omega_D^q) \tag{2.7}$$

is an isomorphism if $p + q < n - 1$ and is injective if $p + q = n - 1$.

If we set $L_D^{-1} = \mathcal{O}_D \otimes \mathcal{J}_D$ then L_D is a very ample invertible sheaf on D, being associated with the hyperplane section of D. By an induction hypothesis, we thus have

$$H^p(X,\Omega_D^{q-1} \otimes L_D^{-1}) = 0 \tag{2.8}$$

if $p + q < n$. The exact cohomology sequence of (2.6) (which we write as $0 \longrightarrow \Omega_D^{q-1} \otimes L_D^{-1} \xrightarrow{df} \Omega_X^q \otimes \mathcal{O}_D \xrightarrow{res} \Omega_D^q \longrightarrow 0$) now yields

$$H^p(X,\Omega_X^q \otimes \mathcal{O}_D) \longrightarrow H^p(X,\Omega_D^q) \tag{2.9}$$

is an isomorphism if $p + q < n - 1$ and is injective if $p + q = n - 1$. Since the restriction map $\Omega_X^q \longrightarrow \Omega_D^q$ factors through $\Omega_X^q \longrightarrow \Omega_X^q \otimes \mathcal{O}_D \longrightarrow \Omega_D^q$ from (2.9) and (2.7) we conclude that the natural map

$$H^p(X,\Omega_X^q) \longrightarrow H^p(X,\Omega_X^q \otimes \mathcal{O}_D) \tag{2.10}$$

is an isomorphism if $p + q < n - 1$. Finally we look at the cohomology sequence of

$$0 \longrightarrow \Omega_X^q \otimes L^{-1} \longrightarrow \Omega_X^q \longrightarrow \Omega_X^q \otimes \mathcal{O}_D \longrightarrow 0$$

and deduce from (2.10) that

$$H^p(X,\Omega_X^q \otimes L^{-1}) = 0$$

if $p+q < n-1$. The last piece of the cohomology sequence for $p+q = n-1$ yields

$$0 \longrightarrow H^p(X,\Omega_X^q \otimes L^{-1}) \longrightarrow H^p(X,\Omega_X^q) \longrightarrow H^p(X,\Omega_X^q \otimes \mathcal{O}_D)$$

and the last arrow is injective by (2.7) and (2.9). It follows that $H^p(X,\Omega_X^q \otimes L^{-1}) = 0$ if $p + q = n - 1$, and the proof of $\underline{C}'$ in case L is very ample is completed by induction on $n = \dim X$.

In the general case in which L is ample but not very ample we use the method of branched coverings, already introduced by Mumford in this type of questions.

Say L^m is very ample and let $D \in |L^m|$ be a smooth divisor of zeros of a section of L^m. With respect to a suitable affine covering $\{U_i\}$ of X let L be determined by the 1-cocycle $\{g_{ij}, U_i \cap U_j\}$ and let $\{f_i, U_i\}$ be a system of local equations for D, hence $f_i = g_{ij}^m f_j$ on $U_i \cap U_j$. We consider now the variety $X' \subset \mathbf{A}^1 \times X$ defined by

$$\begin{cases} z_i^m = f_i & \text{on } \mathbf{A}^1 \times U_i \\ z_i/z_j = g_{ij} & \text{on } \mathbf{A}^1 \times U_i \cap \mathbf{A}^1 \times U_j . \end{cases}$$

First of all, X' is a smooth projective variety, since at a singular point we have

$$\begin{cases} z_i^m - f_i = 0 \\ d(z_i^m - f_i) = 0 \end{cases}$$

hence $z_i = 0$, $f_i = 0$, $df_i = 0$ and $f_i = 0$ is not a smooth divisor, contradicting our assumptions.

Let $\pi : X' \longrightarrow X$ be the covering map. Then $L' = \pi^* L$ is ample. If we consider $\pi^* f$ then locally $\pi^* f_i = f_i = z_i^m$ and $z_i = g_{ij} z_j$, hence $z = \{z_i\}$ is a section of $\pi^* L$ satisfying $z^m = \pi^* f$. The section z defines a smooth divisor D' on X', as one can see by taking local coordinates $x_1, \ldots, x_n$ on X such that $f_i = x_1$, and now $z, x_2, \ldots, x_n$ are corresponding local coordinates on X'. Finally D' is very ample, because

(i) D' belongs to an ample system

(ii) D' is smooth

(iii) $\dim X' \geqq 2$

(iv) L^m is very ample on X;

however, we will not prove this fact here. If L' is the invertible sheaf $L' = \pi^* L$, we see that by the special case considered before we have

$$H^p(X',\Omega^q_{X'}\otimes L'^{-1}) = 0$$

if $p+q<n$. Everything now follows from the fact that $H^p(X,\Omega^q_X\otimes L^{-1})$ is a direct factor of the previous group. To show this, we have to produce a splitting of

$$\Omega^q_X\otimes L^{-1}\hookrightarrow \pi_*(\Omega^q_{X'}\otimes\pi^*L^{-1})$$

which in turn will follow if and only if

$$\Omega^q_X \hookrightarrow \pi_*\Omega^q_{X'}$$

splits. In order to exhibit the required splitting, we use a trace map. Say U is open in X and let

$$\omega\in\Gamma(\pi^{-1}(U),\Omega^q_{X'}\otimes L'^{-1}) .$$

Let $\varepsilon : X'\longrightarrow X'$ be the automorphism of X' determined by $z_i\longrightarrow \varepsilon z_i$ where $\varepsilon^m = 1$ and let

$$\tilde{\omega} = \sum_{\varepsilon}\varepsilon^*\omega ;$$

there is a unique element

$$\mathrm{Tr}(\omega)\in\Gamma(U,\Omega^q_X\otimes L^{-1})$$

such that

$$\pi^*(\mathrm{Tr}\ \omega) = \tilde{\omega} ,$$

and finally

$$\omega\longmapsto\frac{1}{m}\mathrm{Tr}\ \omega$$

is the required splitting. To see this we have to check that $\eta = \frac{1}{m}\mathrm{Tr}(\pi^*\eta)$, which is easily done with a computation in local coordinates.

We conclude with the remark that the condition char $= 0$ has been used twice: the first time, to prove (2.7) and the second time, allowing division by the integer m to define the splitting. It is, however, the failure of (2.7) already when $\dim X = 2$ which is responsible for the failure of Kodaira's vanishing theorem in char $\neq 0$.

Now we analyze in more detail the situation in case $\dim X = 2$. Here the crucial cohomology group to study is $H^1(X,L^{-1})$. If there is a divisor $D\in|L|$, s is a section of L with div $(s) = D$, we have an exact sequence

$$(2.11) \qquad 0 \longrightarrow \mathcal{I}_D \xrightarrow{s} \mathcal{O}_X \longrightarrow \mathcal{O}_D \longrightarrow 0 .$$

It is clear that $H^0(X, \mathcal{I}_D) = 0$ (the only regular functions on X are the constants) hence the cohomology sequence of (2.11) yields

$$(2.12) \qquad \dim H^1(X, L^{-1}) = -1 + h^0(\mathcal{O}_D) + \dim \ker\{H^1(X, \mathcal{O}_X) \longrightarrow H^1(D, \mathcal{O}_D)\} .$$

We are led to

Problem 1. Compute $h^0(\mathcal{O}_D)$.

Problem 2. Compute

$$\alpha(D) = \dim \ker\{H^1(X, \mathcal{O}_X) \longrightarrow H^1(D, \mathcal{O}_D)\} .$$

Lemma (Ramanujam). If either $\mathrm{char}(k) = 0$ or the Frobenius cohomology operation F is injective on $H^1(X, \mathcal{O}_X)$, we have

$$\alpha(D) = \alpha(D_{red}) .$$

Proof. We show that if $D_1 < D_2 \leqq 2D_1$ then $\alpha(D_1) = \alpha(D_2)$; the lemma will follow by induction on the multiplicities of the various components of D. If $\mathrm{char}(k) = 0$, we use a truncated exponential sequence as follows. Since the ideal sheaf $\mathcal{I}_{D_1}\mathcal{O}_{D_2}$ is of square 0 (because $D_1 < D_2 \leqq 2D_1$) the map $\varepsilon \longmapsto 1 + \varepsilon$ is a group homomorphism $\mathcal{I}_{D_1}\mathcal{O}_{D_2} \longrightarrow \mathcal{O}^*_{D_2}$ which fits in an exact sequence

$$0 \longrightarrow \mathcal{I}_{D_1}\mathcal{O}_{D_2} \longrightarrow \mathcal{O}^*_{D_2} \longrightarrow \mathcal{O}^*_{D_1} \longrightarrow 0$$

and we get a cohomology sequence

$$H^0(D_1, \mathcal{O}^*_{D_1}) \xrightarrow{\partial} H^1(D_2, \mathcal{I}_{D_1}\mathcal{O}_{D_2}) \xrightarrow{\rho} \mathrm{Pic}(D_2) \longrightarrow \mathrm{Pic}(D_1) .$$

We shall show that the kernel of $\mathrm{Pic}(D_2) \longrightarrow \mathrm{Pic}(D_1)$ has no torsion. In fact, say $\alpha \in \ker\{\mathrm{Pic}(D_2 \longrightarrow \mathrm{Pic}(D_1)\}$ and $m\alpha = 0$ (we use additive notation except for $H^0(D_1, \mathcal{O}^*_{D_1})$). There is $\beta \in H^1(D_2, \mathcal{I}_{D_1}\mathcal{O}_{D_2})$ such that

$$\rho(\beta) = \alpha ,$$

and now

$$\rho(m\beta) = m\alpha = 0$$

hence

$$m\beta \in \partial H^0(D_1, \mathcal{O}^*_{D_1}) .$$

If $\gamma \in H^0(D_1, \mathcal{O}^*_{D_1})$ is such that

$$\partial\gamma = m\beta ,$$

then $\partial(\gamma^{1/m}) = \beta$ provided we can take an m-th root of γ with $\gamma^{1/m} \in H^0(D_1, \mathcal{O}^*_{D_1})$. In this case however

$$\beta \in \partial H^0(D_1, \mathcal{O}^*_{D_1}) = \ker \rho$$

hence $\alpha = \rho(\beta) = 0$ and there is no torsion in $\ker\{\mathrm{Pic}(D_2) \longrightarrow \mathrm{Pic}(D_1)\}$.

Now γ, being a global section of $\mathcal{O}^*_{D_1}$, is written uniquely as

$$\gamma = \gamma_0 + \varepsilon$$

where γ_0 is locally constant and non-zero on D_1, while ε is nilpotent of order $N \leqq$ max multiplicity of components of D_1. In order to take m-th roots, we may suppose $\gamma_0 = 1$ and, as we are in char = 0, we may use the binomial series (ε is nilpotent)

$$(1 + \varepsilon)^{1/m} = \sum_{j=0}^{N-1} \binom{1/m}{j} \varepsilon^j \in H^0(D_1, \mathcal{O}^*_{D_1}).$$

Now look at

$$\begin{array}{ccc} \mathrm{Pic}^0(X) & \xrightarrow{\alpha} & \mathrm{Pic}(D_2) \\ & \searrow^{\beta} & \downarrow \mathrm{res} \\ & & \mathrm{Pic}(D_1) \end{array}$$

and let

$$K(D_i) = \ker\{\mathrm{Pic}^0(X) \longrightarrow \mathrm{Pic}(D_i)\} .$$

Since $\ker\{\mathrm{Pic}(D_2) \longrightarrow \mathrm{Pic}(D_1)\}$ has no torsion, we also see that $K(D_1)/K(D_2)$ has no torsion either. If A_i is the connected component at 0 of $K(D_i)$ then A_i is a subgroupscheme of the abelian variety $\mathrm{Pic}^0(X)$ and, as we are in char = 0, A_i is reduced and is an abelian variety by Cartier's theorem. Now A_1/A_2 has only torsion of finite order, since $K(D_i)/A_i$ is a finite group and $K(D_1)/K(D_2)$ has no torsion, hence A_1/A_2 is a finite group and $\dim A_1 = \dim A_2$. Finally, as we are in char = 0, we see that $\ker\{H^1(X, \mathcal{O}_X) \longrightarrow H^1(D_i, \mathcal{O}_{D_i})\}$ is the tangent space at 0 of A_i (recall that $K(D_i)$ is reduced, being a groupscheme in char $\neq 0$) hence $\alpha(D_i) = \dim A_i$ and $\alpha(D_1) = \alpha(D_2)$.

If $\mathrm{char}(k) = p$ we proceed in a different and much simpler way. We have the exact sequence

$$0 \longrightarrow \mathcal{J}_{D_1}\mathcal{O}_{D_2} \longrightarrow \mathcal{O}_{D_2} \longrightarrow \mathcal{O}_{D_1} \longrightarrow 0$$

hence a cohomology sequence

$$H^1(X, \mathcal{J}_{D_1}\mathcal{O}_{D_2}) \longrightarrow H^1(D_2, \mathcal{O}_{D_2}) \longrightarrow H^1(D_1, \mathcal{O}_{D_1}) .$$

If we apply the Frobenius cohomology operation F to this sequence we obtain the exact sequence

$$0 \longrightarrow FH^1(D_2, \mathcal{O}_{D_2}) \longrightarrow FH^1(D_1, \mathcal{O}_{D_1})$$

because $FH^1(X, \mathcal{J}_{D_1}\mathcal{O}_{D_2}) = 0$. In fact, F acts on $H^1(X, \mathcal{J}_{D_1}\mathcal{O}_{D_2})$ by raising cocycles to the p-th power, and now $p \geqq 2$ while $\mathcal{J}_{D_1}\mathcal{O}_{D_2}$ is of square 0. Thus we have proved that F kills the kernel of $H^1(D_2, \mathcal{O}_{D_2}) \longrightarrow H^1(D_1, \mathcal{O}_{D_1})$.

Now look at

$$\begin{array}{ccc} H^1(X, \mathcal{O}_X) & \xrightarrow{\alpha} & H^1(D_2, \mathcal{O}_{D_2}) \\ & \overset{\beta}{\searrow} & \downarrow \mathrm{res} \\ & & H^1(D_1, \mathcal{O}_{D_1}) \end{array} .$$

If u is an element of $H^1(X, \mathcal{O}_X)$ with $\alpha(u) \neq 0$ and $\beta(u) = 0$, then $\alpha(u) \in \ker\{H^1(D_2, \mathcal{O}_{D_2}) \longrightarrow H^1(D_1, \mathcal{O}_{D_1})\}$ and $F\alpha(u) = 0$ by our previous remark, hence $\alpha(Fu) = 0$ while $\alpha(u) \neq 0$. Since $F \ker(\alpha) \subset \ker(\alpha)$, we see that F cannot be injective on $H^1(X, \mathcal{O}_X)$. This completes the proof of Ramanujam's lemma.

The next result is very important and useful in dealing with curves on a surface.

Proposition 1. Let $C > 0$ be an effective divisor on a smooth surface X and let L be an invertible sheaf on C with $H^0(C, L) \neq 0$.

Then there is a decomposition

$$C = C_1 + C_2, \ C_1 \geqq 0, \ C_2 > 0$$

such that

$$(C_1 \cdot C_2) \leqq \deg_{C_2}(L \otimes \mathcal{O}_{C_2}) .$$

Moreover, this will hold for every C_1 such that there is a section s of L vanishing on C_1 but not on any C' with $C_1 < C' < C$.

<u>Proof.</u> Let $s \in H^0(C,L)$ be a non-zero section of L. If the restriction of s to an irreducible component of C is never identically 0, then the degree of L on this component is non-negative and a fortiori $\deg_C L \geqq 0$. Now suppose that $0 < C_1 < C$ is a <u>maximal</u> divisor such that s vanishes on C_1. Then we have two exact sequences of sheaves

$$0 \longrightarrow \mathcal{O}_{C_2} \xrightarrow{s} L \longrightarrow L/s\mathcal{O}_C \longrightarrow 0$$

$$0 \longrightarrow F \longrightarrow L/s\mathcal{O}_C \longrightarrow L \otimes \mathcal{O}_{C_1} \longrightarrow 0$$

where F is supported at finitely many points. Taking Chern classes on X we find

$$c(L) = c(\mathcal{O}_{C_2})\, c(L/s\mathcal{O}_C) ,$$

$$= c(\mathcal{O}_{C_2})\, c(L \otimes \mathcal{O}_{C_1})\, c(F)$$

whence the equation

$$\underline{\underline{1}} + C + (C^2 - \deg_C L)$$

$$= (\underline{\underline{1}} + C_2 + C_2^2)(\underline{\underline{1}} - \text{length } F)(\underline{\underline{1}} + C_1 + C_1^2 - \deg_{C_1} L)$$

in the Chow ring of X. The equation in degree 2 is simply

$$(C_1 \cdot C_2) + \text{length } F = \deg_{C_2} L$$

and Proposition 1 is proven.

<u>Definition.</u> An effective divisor D on a non-singular surface X is numerically connected if for every decomposition $D = D_1 + D_2$, $D_i > 0$ we have $(D_1 \cdot D_2) > 0$.

Now we can solve Problem 1 as follows.

<u>Proposition</u> 2. If D is numerically connected, then $h^0(\mathcal{O}_D) = 1$.

<u>Proof.</u> We apply Proposition 1 with $C = D$, $L = \mathcal{O}_D$. Now $\deg_{C_2}(\mathcal{O}_{C_2}) = 0$ for all C_2 and if $D = C_1 + C_2$ with $C_i > 0$ we cannot have $(C_1 \cdot C_2) \leqq 0$. It follows that, by the last clause of Proposition 1, we

must have $C_1 = 0$ hence no section of $\mathcal{O}_D$ can vanish on components of D. This shows that $H^0(D,\mathcal{O}_D)$ consists only of constants and the required conclusion follows.

Now we prove

Ramanujam's vanishing theorem. Assume that $\mathrm{char}(k) = 0$ or the Frobenius cohomology operation is injective on $H^1(X,\mathcal{O}_X)$. Let D be a numerically connected divisor on X and let $L = \mathcal{O}_X(D)$.

Suppose also that for large n the linear system $|L^n|$ is not composite of a pencil and has no fixed components. Then we have

$$H^1(X,L^{-1}) = 0 .$$

Proof. The assumptions about L imply that there is $N > 0$ such that $|L^N|$ has no base points and determines a birational map $\Phi : X \longrightarrow Y$ where Y is a normal surface. We shall show that, taking a larger N if necessary, we have $H^1(X,L^{-N}) = 0$, hence (2.12) yields $\alpha(C) = 0$ for every $C \in |L^N|$. Since $ND \in |L^N|$, we have $\alpha(ND) = 0$ and now Ramanujam's lemma shows that $\alpha(D) = 0$. Using (2.12) again we get

$$\begin{aligned} h^1(L^{-1}) &= -1 + h^0(\mathcal{O}_D) + \alpha(D) \\ &= -1 + h^0(\mathcal{O}_D) = 0, \end{aligned}$$

by Proposition 2, and our theorem is proven.

It remains to check the vanishing of $H^1(X,L^{-N})$ for sufficiently large N. If L is ample, this is a consequence of the Enriques-Severi-Zariski-Serre theorem. In the general case we proceed as follows. Let $\Phi : X \longrightarrow Y$ be as before. We want to show that $H^1(X,\Omega_X^2 \otimes L^{Nn}) = 0$ for all large n and the required result will be a consequence of Serre's duality. Now the Leray spectral sequence for the map Φ shows that $H^1(X,\Omega_X^2 \otimes L^{Nn}) = 0$ if $H^1(Y,\Phi_*(\Omega_X^2 \otimes L^{Nn})) = 0$ and if $R^1\Phi_*(\Omega_X^2 \otimes L^{Nn}) = 0$. The vanishing of the H^1 for large n is a consequence of the Enriques-Severi-Zariski-Serre theorem, because

$$H^1(Y,\Phi_*(\Omega_X^2 \otimes L^{Nn})) = H^1(Y,\Phi_*(\Omega_X^2) \otimes \mathcal{O}_Y(n)) .$$

In order to check the vanishing of $R^1\Phi_*(\Omega_X^2 \otimes L^{Nn})$ we note that, since this sheaf is concentrated at the finitely many points $P \in Y$ such that $\Phi^{-1}(P)$ has dimension 1, it is sufficient to prove that $(R^1\Phi_*(\Omega_X^2 \otimes L^{Nn}))_P = 0$. By Grothendieck's "Théorème fondamentale des morphismes propres" it is sufficient to prove

$$H^1(D,(\Omega_X^2 \otimes L^{Nn})/\mathcal{J}_D\Omega_X^2) = 0$$

for all effective divisors D with support $\Phi^{-1}(P)$. Now the dualizing sheaf on D is $\mathcal{J}_D^{-1}\Omega_X^2/\Omega_X^2$ hence by duality we have to show that

$$H^0(D,L^{-Nn}\otimes \mathcal{J}_D^{-1}/\mathcal{O}_X) = 0 .$$

By Proposition 1, if this last group were non-zero, we could write $D = D_1 + D_2$, $D_1 \geqq 0$, $D_2 > 0$ and get

$$\begin{aligned}(D_1 \cdot D_2) &\leqq \deg_{D_2}((L^{-Nn}\otimes \mathcal{J}_D^{-1}/\mathcal{O}_X)\otimes \mathcal{O}_{D_2})\\ &= (D\cdot D_2) - Nn(L\cdot D_2)\end{aligned}$$

whence

$$Nn(L\cdot D_2) \leqq (D_2^2) .$$

Now $(L\cdot D_2) = 0$, because Φ contracts D to a point, and $(D_2^2) < 0$ because the intersection quadratic form of a contracted divisor is negative definite; this contradiction proves what we wanted.

Finally we mention the following very precise result.

<u>Theorem.</u> Let L be an invertible sheaf on X, $\mathrm{char}(k) = 0$, and suppose

(i) $(L^2) > 0$

(ii) $(L\cdot C) \geqq 0$ for all $C > 0$.

Then we have $H^i(X,L^{-1}) = 0$ for $i < 2$.

Conversely, assume $(L^2) > 0$ and that $H^i(X,L^{-n}) = 0$ for $i < 2$ and n sufficiently large. Then we have $(L\cdot C) \geqq 0$ for all $C > 0$.

This theorem, due to Ramanujam, gives a necessary and sufficient condition for the vanishing of $H^1(X,L^{-1})$ for invertible sheaves with $(L^2) > 0$, in terms of numerical intersection properties of L.

III. The completeness of the characteristic system. The idea of characteristic system is intimately related to the concept of deformations of a curve on a surface. Intuitively speaking, if $\{C_\alpha\}$ is a continuous family of curves on X, it defines a section of the normal bundle N of C_0 in X as follows. The normal bundle of C_0 in X is nearly isomorphic with a neighborhood of C_0 in X, and for small α the curve C_α is a section of this neighborhood; as $\alpha \longrightarrow 0$, the curves C_α will be more and more identified with a section of the normal bundle N of C_0. In terms of divisors of C_0, $C_\alpha \cap C_0$ will be a certain divisor γ_α on C_0 and $\gamma_0 = \lim_{\alpha\to 0} \gamma_\alpha$ is the divisor of the section of the normal bundle of C_0. If S is a family of deformations of C_0, what we have is a characteristic map

$$\left\{\begin{matrix}\text{Tangent space}\\ \text{to } S \text{ at } \alpha = 0\end{matrix}\right\} \xrightarrow{\rho} H^0(C_0,N) .$$

The fundamental question is now whether ρ is bijective and, even better, whether $\dim S = \dim H^0(C_0,N)$.

In order to put this in rigorous form, let $I = \operatorname{Spec} k[\varepsilon]/\varepsilon^2$ and let D be a curve on X. One defines the normal sheaf of D in X by

$$N_D = \mathcal{O}_X(D)/\mathcal{O}_X ;$$

if D is non-singular, it is also the sheaf of germs of sections of the normal bundle of D in X.

Proposition. The set of curves $\mathcal{D} \subset X \times I$ which extend $D \subset X$ is naturally isomorphic with $H^0(X,N_D)$.

Proof. Let $\{f_i,U_i\}$ define D. If $\mathcal{D}$ extends D, local equations for $\mathcal{D}$ are given by $\{f_i + \varepsilon g_i,U_i\}$ where $\varepsilon^2 = 0$ and $(f_i + \varepsilon g_i) = (\text{unit})(f_j + \varepsilon g_j)$ on $U_i \cap U_j$. Now let

$$(f_i + \varepsilon g_i) = (a_{ij} + \varepsilon b_{ij})(f_j + \varepsilon g_j)$$

on $U_i \cap U_j$; this gives

$$f_i = a_{ij}f_j$$
$$g_i = a_{ij}g_j + b_{ij}f_j$$

hence

$$g_i \equiv a_{ij}g_j \bmod (f_j)$$

which shows that $\{g_i \bmod f_i, U_i\}$ transforms by means of the 1-cocycle $\{a_{ij}, U_i \cap U_j\}$, i.e. is a section of the sheaf $\mathcal{O}(D)/\mathcal{J}_D \cong N_D$. It is also clear that every section of N_D gives rise in this way to a unique $\mathcal{D} \subset X \times I$ and our result follows.

Corollary. Given a family of curves $\mathcal{D} \subset X \times S$ and a closed point $s \in S$ there is a canonical homomorphism

$$\left\{\begin{matrix}\text{Zariski tangent} \\ \text{space } T_s \text{ to } S \text{ at } s\end{matrix}\right\} \xrightarrow{\rho} H^0(X, N_D) .$$

Moreover, if $\mathcal{D} \subset X \times S$ is a universal family, ρ is an isomorphism.

The proof of the corollary is obtained by noting that given $t \in T_s$, there is a canonical morphism $f : I \longrightarrow S$ with image s. By base extension to I, we obtain $\mathcal{D}_f \subset X \times I$ which extends D_s and apply the previous proposition. Linearity of ρ can be checked directly.

The above corollary shows that if S is the universal family of deformations of D, then

$$\dim\left\{\begin{matrix}\text{Zariski tangent} \\ \text{space to } S \text{ at } D\end{matrix}\right\} = h^0(N_D)$$

and the problem of deformation becomes that of the smoothness of S at D. The fundamental theorem is the following.

Theorem (Grothendieck-Cartier). If $H^1(X, \mathcal{O}(D)) = 0$, the following are equivalent:

(i) S is non-singular at D

(ii) S is reduced at D

(iii) $\mathrm{Pic}^0(X)$ is reduced

(iv) $\mathrm{Pic}^0(X)$ is non-singular.

Moreover, $\mathrm{Pic}^0(X)$ is reduced if $\mathrm{char}(k) = 0$ (Cartier) or if $p_g = 0$ (Mumford).

We shall not prove this result here and refer to Mumford's book for proof.

IV. Elliptic or quasi-elliptic fibrations.

We are interested in surfaces X containing a one-parameter family of elliptic curves or more generally curves of genus 1. By blowing up the base points of the family, taking a Stein factorization and blowing down exceptional curves of the first kind which appear as components of a fibre, we are led to study

Relatively minimal fibrations. A fibration $f : X \longrightarrow B$ of a smooth surface over a non-singular curve B, with $f_*\mathcal{O}_X = \mathcal{O}_B$, with all fibres of arithmetic genus 1 and such that no exceptional curve of the first kind is a component of a fibre, is called a relatively minimal elliptic fibration or quasi-elliptic fibration.

Since the function field $k(X)$ is separable over $k(B)$, almost all fibres are generically smooth, and since we deal with the case in which $f_*\mathcal{O}_X = \mathcal{O}_B$ we see that almost all fibres are irreducible. It follows that they can be only of the following kinds:

(a) non-singular elliptic

(b) rational with a node

(c) rational with a cusp.

Now case (b) cannot occur, and case (c) may occur only in characteristics 2 and 3. We shall distinguish between (a) and (c) by talking about elliptic fibrations and quasi-elliptic fibrations respectively.

A curve $D = \Sigma\, n_i E_i$ on X is said to be of canonical type if $(K \cdot E_i) = (D \cdot E_i) = 0$ for all components E_i. If D is not a multiple of another curve of canonical type, D is said to be indecomposable. Now every fibre of $f : X \longrightarrow B$ is of canonical type. In fact, if C is a general fibre we have

$$(D \cdot E_i) = (C \cdot E_i) = 0$$

because C and the fibre $D = \Sigma\, n_i E_i$ are disjoint; to prove that $(K \cdot E_i) = 0$ too, we note that if D is reducible then $(E_i^2) \leqq 0$ because

$$n_i(E_i^2) = (D\cdot E_i) - \sum_{j \neq i} n_j(E_i\cdot E_j)$$

$$= - \sum_{j\neq i} n_j(E_i\cdot E_j) \leqq 0$$

as E_i and E_j are distinct for i, j, and in fact $(E_i^2) < 0$ if D_{red} has two or more components, because D_{red} is connected and hence $(E_i\cdot E_j) > 0$ for at least one index $j \neq i$. Now we cannot have $(K\cdot E_i) < 0$, for otherwise

$$-2 \leqq 2p(E_i) - 2 = (K\cdot E_i) + (E_i^2) \leqq -2$$

and we must have $(K\cdot E_i) = -1$, $(E_i^2) = -1$ and E_i would be an exceptional curve of the first kind, against the hypothesis that our fibration is relatively minimal. It follows that $(K\cdot E_i) \geqq 0$ for all components E_i and finally the assumption $p(D) = 1$ together with $(D^2) = 0$ yields

$$0 = (K\cdot D) + (D^2) = (K\cdot D)$$

$$= \Sigma\, n_i(K\cdot E_i) \geqq 0$$

and all $(K\cdot E_i)$ must be 0.

At finitely many points $b_1,\dots,b_r \in B$ the fibre $f^{-1}(b_\lambda)$ is multiple, i.e.

$$f^{-1}(b_\lambda) = m_\lambda\, P_\lambda$$

with $m_\lambda \geqq 2$ and P_λ indecomposable of canonical type.

We need:

<u>Lemma</u>. Let $D = \Sigma\, n_i\, C_i$ be an effective divisor on X with each C_i irreducible. Assume that $(C_i\cdot D) \leqq 0$ for all i and that D is connected. Then every divisor $Z = \Sigma\, m_i C_i$ satisfies $Z^2 \leqq 0$ and equality holds if and only if $D^2 = 0$ and $Z = \lambda D$, $\lambda \in \mathbb{Q}$.

<u>Proof</u>. Write $x_i = m_i/n_i$. We have

$$\begin{aligned} Z^2 &= \Sigma\, x_i x_j\, n_i n_j (C_i\cdot C_j) \\ &\leqq \Sigma\, x_i^2 n_i^2 (C_i\cdot C_i) + \sum_{i\neq j} \tfrac{1}{2}(x_i^2 + x_j^2)\, n_i n_j (C_i\cdot C_j) \\ &= \Sigma\, x_i^2 n_i (C_i\cdot D) \leqq 0 . \end{aligned}$$

If equality holds everywhere we have either $x_i = x_j$ or $(C_i\cdot C_j) = 0$ for all

i, j and since D is connected, x_i is constant, i.e. $m_i = \lambda n_i$, $\lambda \in \mathbf{Q}$.

Corollary. If D is indecomposable of canonical type, then D is numerically connected, hence $\dim H^0(D, \mathcal{O}_D) = 1$.

Proof. Let $D = D_1 + D_2$, $D_i > 0$. Now $2(D_1 \cdot D_2) = (D^2) - (D_1^2) - (D_2^2)$

$$= -(D_1^2) - (D_2^2) > 0$$

by the previous lemma and the fact that D is indecomposable, hence D_1, D_2 cannot be proportional to D. The last clause of the corollary now follows from Proposition 2 of Ch. II.

We return to the question of the nature of the fibres of f. We have

$$R^1 f_* \mathcal{O}_X = L \oplus T \tag{3.1}$$

where L is an invertible sheaf on B and T is supported at the points $b \in B$ with

$$\dim H^0(f^{-1}(b), \mathcal{O}_{f^{-1}(b)}) \geqq 2 ; \tag{3.2}$$

in fact, by EGA III 7.8, $R^1 f_* \mathcal{O}_X$ is locally free at b if and only if $\mathcal{O}_X$ is cohomologically flat at b in dimension 0. In view of the preceding corollary, the fibre $f^{-1}(b)$ must be decomposable and, in particular, is a multiple fibre. Since not every multiple fibre verifies (3.2), this suggests:

Definition. The multiple fibres over supp T are called wild fibres.

A fundamental result in the theory of elliptic or quasi-elliptic surfaces determines the canonical divisor on X in terms of the fibres of f. We have

Theorem 1. Let $f : X \longrightarrow B$ be a relatively minimal elliptic or quasi-elliptic fibration and let $R^1 f_* \mathcal{O}_X = L \oplus T$. Then

$$\omega_X = f^*(L^{-1} \otimes \omega_B) \otimes \mathcal{O}(\Sigma\, a_\lambda P_\lambda) \tag{3.3}$$

where

(i) $m_\lambda P_\lambda$ are the multiple fibres

(ii) $0 \leqq a_\lambda < m_\lambda$

(iii) $a_\lambda = m_\lambda - 1$ if P_λ is not wild

(iv) $\deg(L^{-1} \otimes \omega_B) = 2p(B) - 2 + \chi(\mathcal{O}_X) + \text{length } T$

where $p(B)$ is the genus of B.

<u>Proof.</u> For any non-multiple fibre $f^{-1}(y)$ we have

$$\mathcal{O}_{f^{-1}(y)} \otimes \omega_X \cong \omega_{f^{-1}(y)} \cong \mathcal{O}_{f^{-1}(y)} \ ,$$

hence if $y_1,\ldots,y_r$ are distinct points of B the cohomology sequence of

$$0 \longrightarrow \omega_X \longrightarrow \omega_X \otimes \mathcal{O}\left(\sum_{i=1}^{r} f^{-1}(y_i)\right) \longrightarrow \bigoplus_{i=1}^{r} \mathcal{O}_{f^{-1}(y_i)} \longrightarrow 0$$

yields

$$\dim\left|\omega_X \otimes \mathcal{O}\left(\sum_{i=1}^{r} f^{-1}(y_i)\right)\right| \geqq 0$$

provided we take r large enough. If D is in the linear system above, we have $(D\cdot f^{-1}(y)) = 0$ therefore D cannot have components transversal to fibres and we can write

$$K_X \equiv (\text{sum of fibres}) + \Delta$$

where $\Delta \geqq 0$ is contained in a sum of fibres and does not contain whole fibres of f. If Δ_ν is a connected component of Δ, by the previous lemma we have $\Delta_\nu^2 \leqq 0$ with equality if and only if Δ_ν is a rational submultiple of the fibre containing Δ_ν. It follows that

$$(K_X^2) = (\Delta^2) = \Sigma\, (\Delta_\nu^2) \leqq 0$$

and each $(\Delta_\nu^2) \leqq 0$. Moreover, we have $0 = (K_X\cdot\Delta_\nu) = (\Delta\cdot\Delta_\nu) = (\Delta_\nu^2)$, hence $(\Delta_\nu^2) = 0$, the fibre C_ν containing Δ_ν is multiple, $C_\nu = m_\nu P_\nu$, and $\Delta_\nu = a_\nu P_\nu$ for some integer $0 \leqq a_\nu < m_\nu$. We have also shown that we must have $(K_X^2) = 0$.

We have proved that

$$\omega_X = f^*\mathcal{O}_B(A) \otimes \mathcal{O}(\Sigma\, a_\lambda P_\lambda)$$

for a suitable $A \in \operatorname{div}(B)$ and we deduce

$$f_*\omega_X = \mathcal{O}_B(A)\ .$$

Now the duality theorem for a map says that

$$\begin{aligned} f_*\omega_X &= \operatorname{Hom}(R^1 f_*\mathcal{O}_X,\ \omega_B) \\ &= L^{-1} \otimes \omega_B \end{aligned}$$

because the dual of a torsion sheaf is 0, and it follows that

$$\omega_X = f^*(L^{-1} \otimes \omega_B) \otimes \mathcal{O}(\Sigma\, a_\lambda P_\lambda)\ .$$

The spectral sequence of the map f yields

$$\chi(\mathcal{O}_X) = \chi(\mathcal{O}_B) - \chi(R^1 f_* \mathcal{O}_X)$$
$$= \chi(\mathcal{O}_B) - \chi(L) - \text{length } T$$
$$= -\deg L - \text{length } T,$$

by the Riemann-Roch theorem on B. Since $\deg \omega_B = 2p(B) - 2$ we obtain statement (iv) of our theorem.

It remains to prove (iii), and this follows from

Proposition 3. Let m_λ, P_λ, a_λ be as in Theorem 1 and let

$$\nu_\lambda = \text{order } (\mathcal{O}_{P_\lambda} \otimes \mathcal{J}_{P_\lambda}^{-1})$$

where $\mathcal{J}_{P_\lambda}$ is the sheaf of ideals of P_λ, be the order of the normal sheaf of P_λ. Then we have:

(i) ν_λ divides m_λ and $a_\lambda + 1$

(ii) $h^0(\mathcal{O}_{(\nu_\lambda+1)P_\lambda}) \geqq 2$, $h^0(\mathcal{O}_{\nu_\lambda P_\lambda}) = 1$

(iii) $h^0(\mathcal{O}_{rP_\lambda})$ is non-decreasing with r.

In particular, if $a_\lambda < m_\lambda - 1$ then $\nu_\lambda < m_\lambda$ and this is equivalent to the multiple fibre $m_\lambda P_\lambda$ being wild.

Proof. In what follows we drop the suffix λ everywhere for simplicity in writing. If $r \geqq s \geqq 1$ the restriction map $\mathcal{O}_{rP} \longrightarrow \mathcal{O}_{sP}$ is surjective, hence $h^1(\mathcal{O}_{rP})$ is non-decreasing with r. Since $\chi(\mathcal{O}_{rP}) = 0$, this proves that $h^0(\mathcal{O}_{rP})$ is non-decreasing too.

We have an isomorphism

$$\mathcal{O}_P \otimes \mathcal{J}^\nu \cong \mathcal{O}_P$$

and via this isomorphism we get an exact sequence

$$0 \longrightarrow \mathcal{O}_P \longrightarrow \mathcal{O}_{(\nu+1)P} \xrightarrow{\text{res}} \mathcal{O}_{\nu P} \longrightarrow 0$$

where res is the restriction. Since constants in $H^0(P, \mathcal{O}_{(\nu+1)P})$ are mapped into constants in $H^0(P, \mathcal{O}_{\nu P})$, the cohomology sequence shows that $\dim H^0(P, \mathcal{O}_{(\nu+1)P}) \geqq 2$. Finally ν divides both m and $a + 1$, because $\mathcal{O}_P \otimes \mathcal{J}^{-m} \cong \mathcal{O}_P$ (trivial) and

$$\mathcal{O}_P \otimes \mathcal{J}^{-a-1} \cong \omega_P = \mathcal{O}_P .$$

In order to check the isomorphism $\omega_P \cong \mathcal{O}_P$, we note that since $\chi(\mathcal{O}_P) = 0$ and

$h^0(\mathcal{O}_P) = 1$, we have $h^0(\omega_P) = h^1(\mathcal{O}_P) = 1$, hence ω_P has a non-trivial section s. Now ω_P has degree 0 on every component of P and P is numerically connected. By Proposition 1 of Ch. II, we see that s cannot vanish identically on any component of P. Also, s cannot have isolated zeros, because ω_P has degree 0 on each component. We deduce that s never vanishes and hence multiplication by s^{-1} yields the required isomorphism $\omega_P \xrightarrow{\sim} \mathcal{O}_P$. This proves Proposition 3.

Corollary. If in addition $h^1(\mathcal{O}_X) \leqq 1$ we have either $a_\lambda + 1 = m_\lambda$ or $\nu_\lambda + a_\lambda + 1 = m_\lambda$.

Proof. Since $\chi(\mathcal{O}_{(\nu+1)P}) = 0$ and $h^0(\mathcal{O}_{(\nu+1)P}) = 0$, using duality we find

$$\dim H^0(P, \omega_{(\nu+1)P}) \geqq 2 .$$

Now the cohomology sequence of

$$0 \longrightarrow \omega_X \longrightarrow \mathcal{J}^{-\nu-1} \otimes \omega_X \longrightarrow \omega_{(\nu+1)P} \longrightarrow 0$$

yields

$$\dim H^0(X, \mathcal{J}^{-\nu-1} \otimes \omega_X) > \dim H^0(X, \omega_X)$$

since $\dim H^1(X, \omega_X) = \dim H^1(X, \mathcal{O}_X) \leqq 1$ by hypothesis. This increase in dimension is possible only if $\nu + a + 1 \geqq m$, or $1 + (a+1)/\nu \geqq m/\nu$. Therefore $(a+1)/\nu = m/\nu$ or $m/\nu - 1$.

In some cases it may happen that there is a multiple fibre mP such that $\nu = 0$, i.e. P has a trivial normal sheaf. By the results of Ch. III we now see that the universal deformation space of P consists of a single point but with Zariski tangent space of dimension 1; although there are non-trivial first-order deformations of P, we find obstructions to extend these deformations to higher order.

The occurrence of wild fibres is a phenomenon peculiar to char. p. In fact Kodaira has shown that in char. 0 we must have $\nu_\lambda = m_\lambda$, $a_\lambda = m_\lambda - 1$ and more generally Raynaud has proved that m_λ/ν_λ is a power of the characteristic p.

Now we shall examine in more detail the case in which $B = \mathbf{P}^1$, the projective line.

If $B = \mathbf{P}^1$, all fibres are rationally equivalent and the canonical bundle formula takes a simpler form. We have

<u>Theorem</u> 2. If $B = \mathbf{P}^1$ then

$$K_X \equiv rf^{-1}(y) + \sum_\lambda a_\lambda P_\lambda$$

where $r = -2 + \chi(\mathcal{O}_X) + \text{length } T$. Moreover

$$r + \sum_\lambda \frac{a_\lambda}{m_\lambda} \geqq 0$$

unless X is rational or ruled and the plurigenera of X are then given by the formula

$$P_n = \max\left(nr + \sum_\lambda \left[\frac{na_\lambda}{m_\lambda}\right] + 1,\ 0\right)$$

where $[\ ]$ denotes the integral part.

<u>Proof</u>. The first formula is obvious, because all fibres are rationally equivalent, i.e. invertible sheaves on $\mathbf{P}^1$ are classified by their degree. The inequality $r + \sum_\lambda \frac{a_\lambda}{m_\lambda} \geqq 0$ also comes from $(K_X \cdot H) \geqq 0$ for all ample H unless X is rational or ruled (Castelnuovo, Enriques, Mumford). Since

$$nK_X \equiv nr\ f^{-1}(y) + \sum_\lambda na_\lambda P_\lambda$$

we have

$$nK_X \equiv \left(nr + \sum_\lambda \left[\frac{na_\lambda}{m_\lambda}\right]\right) f^{-1}(y) + \sum_\lambda b_\lambda(n) P_\lambda$$

where $0 \leqq b_\lambda(n) < m_\lambda$, hence

$$\begin{aligned} &\dim H^0(X, \mathcal{O}(nK_X)) \\ &= \dim H^0\left(B, \mathcal{O}_B\left(nr + \sum_\lambda \left[\frac{na_\lambda}{m_\lambda}\right]\right)\right) \\ &= nr + \sum_\lambda \left[\frac{na_\lambda}{m_\lambda}\right] + 1 \end{aligned}$$

provided $nr + \sum_\lambda \left[\frac{na_\lambda}{m_\lambda}\right] \geqq -1$, and our result follows.

In any case, we have that P_n cannot grow more than linearly with n. If one defines the Kodaira dimension to be

$$\varkappa = \text{tr. deg.}_k \bigoplus_{n=0}^{\infty} H^0(X, \mathcal{O}(nK)) - 1$$

we have

$$\varkappa = -1 \Longleftrightarrow \text{all plurigenera } P_n \text{ vanish}$$

$$\varkappa = 0 \Longleftrightarrow P_n \leqq 1 \text{ for all } n, \text{ and } P_n \neq 0 \text{ for some } n$$

$$\varkappa = 1 \Longleftrightarrow P_n \sim \gamma n, \text{ some } \gamma > 0$$

$$\varkappa = 2 \Longleftrightarrow P_n \sim \gamma n^2, \text{ some } \gamma > 0 .$$

If $\varkappa = -1$, then X is rational or ruled (Castelnuovo, Enriques, Mumford), while if $\varkappa = 1$ then X is elliptic or quasi-elliptic (Enriques, Mumford). If $\varkappa = 0$, the situation is more complicated, since X need not be elliptic. However, if X is elliptic the canonical bundle formula and the Riemann-Roch theorem on B shows that $P_n \sim \gamma n$ where $\gamma \geqq 0$

$$\gamma = r + \Sigma \frac{a_\lambda}{m_\lambda} .$$

We deduce that

<u>Theorem</u> 3. Let X be an elliptic or quasi-elliptic surface with $\varkappa = 0$ or 1. Then if $f : X \longrightarrow B$ is a relatively minimal fibration we have that

$$2p(B) - 2 + \text{length } T + \sum_\lambda \frac{a_\lambda}{m_\lambda}$$

does not depend on the fibration f.

Theorem 3 is meaningful, since some surfaces may have more than one fibration.

Going back to surfaces with $\varkappa = 0$ or 1 we see that minimal models of them all have $(K_X^2) = 0$. In fact, if $(K_X^2) > 0$ then $\dim|nK_X| + \dim|-(n-1)K_X| \longrightarrow \infty$ as $n \longrightarrow \infty$, by the Riemann-Roch theorem, quadratically with n. This implies that either $P_n = 0$ for all n (hence $\varkappa = -1$) or $P_n \sim \gamma n^2$, $\gamma > 0$ (hence $\varkappa = 2$), so that we deduce that if $\varkappa = 0$ or 1 then $(K_X^2) \leqq 0$. If $(K_X^2) < 0$ and X is minimal then, as proved by Mumford in all characteristics, we have that X is ruled and $\varkappa = -1$. This being done, the Riemann-Roch theorem for the sheaf $\mathcal{O}_X$ yields

$$\begin{aligned} &12(h^0(\mathcal{O}_X) - h^1(\mathcal{O}_X) + h^2(\mathcal{O}_X)) \\ &= c_2[X] = B_0 - B_1 + B_2 - B_3 + B_4 \\ &= 2 - 4q + B_2 \end{aligned}$$

and we obtain the

<u>Basic formula</u>. If $(K_X^2) = 0$ we have

$$10 + 12p_g = 8 \dim H^1(X,\mathcal{O}_X) + 2(2 \dim H^1(X,\mathcal{O}_X) - 2q) + B_2 .$$

It is known that

$$\dim H^1(X,\mathcal{O}_X) = q$$

if $\mathrm{char}(k) = 0$ and more precisely

$$\dim H^1(X,\mathcal{O}_X) \geqq q$$

in any characteristic, with equality if and only if $\mathrm{Pic}(X)$ is reduced; we have also (Bombieri-Mumford)

$$\dim H^1(X,\mathcal{O}_X) \leqq q + p_g .$$

If we give an upper bound for p_g, the basic formula shows that we have only finitely many possibilities for $h^1(\mathcal{O}_X)$, B_2 and $\Delta = 2h^1(\mathcal{O}_X) - 2q$. In particular we have

<u>Table of Invariants if</u> $(K^2) = 0$, $p_g \leqq 1$.

B_2	B_1	c_2	$\chi(\mathcal{O}_X)$	$h^1(\mathcal{O}_X)$	p_g	Δ
22	0	24	2	0	1	0
14	2	12	1	1	1	0
10	0	12	1	$\begin{cases}0\\1\end{cases}$	$\begin{cases}0\\1\end{cases}$	$\begin{cases}0\\2\end{cases}$
6	4	0	0	2	1	0
2	2	0	0	$\begin{cases}1\\2\end{cases}$	$\begin{cases}0\\1\end{cases}$	$\begin{cases}0\\2\end{cases}$

The first four invariants are deformation invariants, since they are determined by the Betti numbers, while the last three are not always deformation invariants if $\mathrm{char}(k) \neq 0$. This suggests that Enriques' classification by means of plurigenera should be replaced in $\mathrm{char}(k) = p$ by means of a classification using the Betti and Chern numbers, together with the Kodaira dimension.

We mention here a new phenomenon of char. p. If char. = 0, then a classical theorem of Castelnuovo shows that if $\chi(\mathcal{O}_X) < 0$ then X is ruled. This is no longer true in char. p, and Raynaud has constructed non-ruled surfaces with $p_g > 0$ and $\chi(\mathcal{O}_X) < 0$. These surfaces admit fibrations $f : X \longrightarrow B$ with fibres of positive arithmetic genus having cusp singularities. The following result of W. Lang shows that this is intimately related with the failure of Kodaira's vanishing theorem in char. p.

Theorem 4. Let $f : X \longrightarrow B$ be a fibration such that all fibres are irreducible and have positive arithmetic genus, admitting a section C. Let $N_C = \mathcal{O}_X(C)/\mathcal{O}_X$ be the normal sheaf on C (and $C \cong B$ since C is a section).

Then if $C^2 > 0$ we have that

(i) $H = \mathcal{O}_X(C) \otimes f^* N_C$ is ample

(ii) $H^1(X, H^{-1}) \neq 0$.

Now we shall consider surfaces with $\kappa = 0$. We cannot have $p_g \geqq 2$, for if s_1, s_2 are linearly independent sections of $\mathcal{O}(K_X)$ then $s_1^i s_2^{n-i}$, $i = 0,1,\dots,n$ are linearly independent sections of $\mathcal{O}(nK_X)$ and $P_n \geqq n+1$ for all n, hence $\kappa \geqq 1$. Thus surfaces with $\kappa = 0$ have invariants according to the previous table. Let X be such a surface and suppose that $f : \tilde{X} \longrightarrow X$ is an étale covering of degree n. We have $\omega_{\tilde{X}} = f^* \omega_X$ and also $H^0(\tilde{X}, \mathcal{O}(mK_{\tilde{X}})) \cong H^0(X, f_* \mathcal{O}(mK_{\tilde{X}})) \cong H^0(X, \mathcal{O}(mnK_X))$, hence $\kappa = 0$ for $\tilde{X}$ too. Now $\chi(\mathcal{O}_{\tilde{X}}) = n\chi(\mathcal{O}_X)$ and $\chi(\mathcal{O}_{\tilde{X}}) \leqq 2$ and from the table of invariants we see that $n = 1$ if $\chi(\mathcal{O}_X) = 2$, $n \leqq 2$ if $\chi(\mathcal{O}_X) = 1$ or $\chi(\mathcal{O}_X) = 0$. This shows at once that the case in which $B_2 = 14$, $B_1 = 2$ cannot occur if $\kappa = 0$ because $B_1 = 2$ implies $\dim \mathrm{Alb}(X) = 1$, hence X has connected étale coverings of arbitrarily high order, obtained by pull-back of étale coverings of Alb(X). This leads to the following preliminary classification of minimal models.

(a) $\kappa = 0$, $B_2 = 22$, $B_1 = 0$. The surface is called a K3 surface.

(b) $\kappa = 0$, $B_2 = 10$, $B_1 = 0$. We call these Enriques' surfaces.

(c) $\kappa = 0$, $B_2 = 6$, $B_1 = 4$. We shall prove that this characterizes abelian surfaces.

(d) $\kappa = 0$, $B_2 = 2$, $B_1 = 2$. These surfaces are called hyperelliptic or quasi-hyperelliptic.

We shall not deal here with surfaces in (d), but mention only the following facts. If X is as in (d), then $f : X \longrightarrow \mathrm{Alb}(X)$ is an elliptic fibration and there exists a second structure $\pi : X \longrightarrow \mathbf{P}^1$ of X as an elliptic or quasi-elliptic surface. For a detailed analysis of these fibrations and the corresponding finer classification we refer to Bombieri-Mumford.

V. Enriques' surfaces in char. 2. In this chapter we study Enriques' surfaces. Classically, they are defined by means of plurigenera: $p_g = 0$, $P_2 = 1$, $q = 0$, $\varkappa = 0$. In our table, we find however two distinct types, one with $p_g = 0$, $q = 0$, $h^1(\mathcal{O}_X) = 0$ and another with $p_g = 1$, $q = 0$ but $h^1(\mathcal{O}_X) = 1$ (hence $\mathrm{Pic}(X)$ is not reduced and $\mathrm{char}(k) \neq 0$). It is reasonable to keep the name Enriques' surfaces for both types, because it can be shown that deformations of classical Enriques' surfaces in char. 2 may lead to surfaces of the second type.

Theorem 4. Non-classical Enriques' surfaces (i.e. with $p_g = 1$) can occur only in char. 2.

Proof. Let X be a non-classical Enriques' surface. Since $\varkappa = 0$, a theorem of Mumford guarantees that $K_X \equiv 0$ (in fact, by the Riemann-Roch theorem we have $\dim|2K_X| + \dim|-K_X| \geqq 0$ and $\dim|2K_X| = 0$ because $\varkappa = 0$ and $\dim|K_X| = p_g - 1 = 0$, hence $\dim|-K_X| \geqq 0$; since both $|K_X|$ and $|-K_X|$ are non-empty, we must have $K_X \equiv 0$). We have $\dim H^1(X,\mathcal{O}_X) = 1$; hence let $\{a_{ij}\} \in Z^1(\mathcal{O}_X)$ be a non-trivial 1-cocycle and consider the $\mathbf{G}_a$-bundle $\pi : W \longrightarrow X$ defined locally as $\mathbf{A}^1 \times U_i$, coordinate z_i on $\mathbf{A}^1$, and glued by

$$z_i = z_j + a_{ij} .$$

There is a nowhere vanishing 2-form ω on X and $\eta = dz_i \wedge \omega$ is a regular 3-form on W, because $da_{ij} \wedge \omega = 0$ identically. Clearly η never vanishes and we conclude that $K_W \equiv 0$. Now consider the Frobenius cohomology operation

F on $H^1(X,\mathcal{O}_X)$. At the cocycle level, we start with $\{a_{ij}\}$ and obtain the new 1-cocycle $\{a_{ij}^p\}$. Since $\{a_{ij}\}$ is non-trivial and $H^1(X,\mathcal{O}_X)$ is one-dimensional, there is a constant $\lambda \in k$ such that $\{a_{ij}^p\}$, $\{a_{ij}\}$ are cohomologous, hence we can write

$$a_{ij}^p = \lambda a_{ij} + b_i - b_j .$$

Now

$$f = z_i^p - \lambda z_i - b_i$$

defines a global function on W, because $z_i = z_j + a_{ij}$ and $a_{ij}^p = \lambda a_{ij} + b_i - b_j$. Let Y be the surface $f = 0$ in W. If $\lambda \neq 0$ we may assume $\lambda = 1$ and $z_i \longmapsto z_i + \alpha$, $\alpha \in \mathbf{F}_p$ shows that the projection $\pi : Y \longrightarrow X$ is an étale p-covering, of Artin-Schreier type with group $\mathbf{F}_p$. Hence Y is a non-singular (connected) surface. If $\lambda = 0$ we cannot have $b_i \in \mathcal{O}_X^p$ for all i, otherwise $\{a_{ij}\}$ would be trivial, hence df is not identically 0. Thus Y is reduced and a complete intersection in a smooth 3-manifold, hence a Gorenstein surface. Now $K_W \equiv 0$ and the normal sheaf of Y in W is also trivial and we see that in any case

$$\omega_Y \cong \mathcal{O}_Y$$

is the trivial sheaf. We deduce that

$$\begin{aligned} \chi(\mathcal{O}_Y) &\leqq h^0(\mathcal{O}_Y) + h^2(\mathcal{O}_Y) \\ &= h^0(\mathcal{O}_Y) + h^0(\omega_Y) = 2 . \end{aligned}$$

On the other hand, $\pi : Y \longrightarrow X$ is finite and flat and $\pi_*\mathcal{O}_Y$ is filtered by subsheaves

$$\mathcal{O}_X \subset [\mathcal{O}_X \oplus z_i\mathcal{O}_X] \subset [\mathcal{O}_X \oplus z_i\mathcal{O}_X \oplus z_i^2\mathcal{O}_X] \\ \cdots \subset \pi_*\mathcal{O}_Y$$

because $z_i - z_j \in \mathcal{O}_X$, and we have p quotients all isomorphic to $\mathcal{O}_X$. Finally

$$\begin{aligned} \chi(Y,\mathcal{O}_Y) &= \chi(X,\pi_*\mathcal{O}_Y) = \Sigma\, \chi(X, \text{ quotient sheaves}) \\ &= p\, \chi(X,\mathcal{O}_X) = p \end{aligned}$$

and $p \leqq 2$ follows from the previous inequality $\chi(\mathcal{O}_Y) \leqq 2$. Hence $p = 2$

as asserted.

We also deduce that $\chi(\mathcal{O}_Y) = 2$, $\omega_Y \cong \mathcal{O}_Y$ and $h^1(\mathcal{O}_Y) = 0$, i.e. if Y is non-singular then Y is a K3-surface while in any case it is K3-like.

Theorem 5. Every Enriques surface is elliptic or quasi-elliptic.

Proof. We shall not deal with the classical case, since the known proofs in char. 0 extend easily to char. p. So assume that X is non-classical.

Everything follows (as in char. 0) from the following result.

Proposition. Let C be an irreducible curve on X with $(C^2) > 0$. Then the linear system $|C|$ contains a reducible divisor D which is not the sum of two non-singular rational curves.

We postpone the proof of this result and go first to the proof of our theorem.

Let $p = \frac{1}{2}(C^2) + 1$ be the smallest genus of irreducible curves with $(C^2) > 0$, and let D be as in the Proposition.

Step 1. If E is an irreducible component of D then $p(E) < p(C)$ hence $p(E) = 0$ or 1.

In fact, let $D = E + D'$. We have $(D\cdot D') = (C\cdot D') \geqq 0$ because C is irreducible and $(C^2) > 0$, hence $(D\cdot E) \leqq (C^2)$. Now $(D\cdot E) = (E^2) + (E\cdot D')$ and D_{red} is connected, hence either $(E\cdot D') > 0$ or $(E^2) < 0$ and in both cases $(E^2) = (D\cdot E) - (E\cdot D') < (C^2)$, i.e. $p(E) < p(C)$.

Step 2. Either X contains a curve of canonical type or $D = \Sigma\, m_i E_i$ where $\Sigma\, m_i \geqq 3$ and the E_i are non-singular rational curves with $(E_i^2) = -2$ and $(E_i\cdot E_j) \leqq 2$ for all i, j.

In fact, if $p(E_i) = 1$ then E_i is of canonical type, hence we may suppose $(E_i^2) = -2$ and E_i non-singular rational. Also $\Sigma\, m_i \geqq 3$ by the previous proposition. Finally, $(E_i\cdot E_j) \leqq 2$. For if $(E_i\cdot E_j) \geqq 3$ then $E_i + E_j$ has self-intersection $\geqq 2$ and the Riemann-Roch theorem yields $\dim|E_i + E_j| \geqq 1$. Note also that the general element of $|E_i + E_j|$ is irreducible. On the other hand

$$|E_i + E_j| + (D - E_i - E_j) \subset |C|$$

and $D - E_i - E_j > 0$, because $\Sigma\, m_i \geqq 3$; this clearly contradicts the result in Step 1.

<u>Final step</u>. X contains a curve of canonical type. Consider $D = \Sigma\, m_i E_i$ as in Step 2. If $(E_i \cdot E_j) = 2$ then $E_i + E_j$ is of canonical type. Hence we may assume $(E_i \cdot E_j) \leqq 1$ for all i, j. Consider the connected graph with vertices E_i and edges connecting E_i, E_j if $(E_i \cdot E_j) = 1$. If this graph contains a complete Dynkin diagram $\tilde{A}_n$, $\tilde{D}_n$, $\tilde{E}_6$, $\tilde{E}_7$, $\tilde{E}_8$ (e.g. $\tilde{A}_n$ is a loop) then by attaching suitable multiplicities to the vertices of this subgraph we get a curve of canonical type. Otherwise the graph itself is a Dynkin diagram and the associated self-intersection quadratic form is negative definite, which contradicts $(D^2) > 0$. Finally X is elliptic or quasi-elliptic since it contains a curve of canonical type.

It remains to prove the Proposition.

Let L be an invertible sheaf on X, $L \not\cong \mathcal{O}_X$ and $|L| \neq \emptyset$. If $s \in \Gamma(L)$ and $C = \mathrm{div}(s)$ we have the exact sequence

$$0 \longrightarrow \mathcal{O}_X \xrightarrow{\ s\ } L \longrightarrow \mathcal{O}_C \otimes L \longrightarrow 0$$

and we define $\Gamma(L)_0$ as the vector subspace of $\Gamma(L)$ consisting of those sections $s \in \Gamma(L)$ such that

$$H^1(X,\mathcal{O}_X) \xrightarrow{\ s\ } H^1(X,L)$$

is the zero map; $|L|_0$ will be the associated (possibly non-complete) linear system.

<u>Lemma</u> 1. Assume that there exists $C \in |L|$ with $h^0(\mathcal{O}_C) = 1$. Then we have:

either

a) $h^1(L) = 0$ and $\dim|L| = \dim|L|_0 = \frac{1}{2}(L^2)$

or

b) $h^1(L) = 1$ and $\dim|L|_0 = \frac{1}{2}(L^2)$.

Moreover, in case b) we have $D \in |L|_0$ if and only if $h^0(\mathcal{O}_D) = 2$, hence every element of $|L|_0$ is reducible and is not the sum of two rational curves.

Proof. The Riemann-Roch theorem yields

$$\dim|L| = \frac{1}{2}(L^2) + h^1(L) .$$

The cohomology sequence of

$$0 \longrightarrow \mathcal{O}_X \xrightarrow{s} L \longrightarrow \omega_D \longrightarrow 0$$

where $D = \mathrm{div}(s)$ gives

$$H^1(X,\mathcal{O}_X) \xrightarrow{s} H^1(X,L) \longrightarrow H^1(D,\omega_D) \longrightarrow H^2(X,\mathcal{O}_X) \longrightarrow 0$$

and now choosing $D = C$, hence with $h^1(\omega_D) = h^0(\mathcal{O}_D) = 1$ we see that $H^1(D,\omega_D) \longrightarrow H^2(X,\mathcal{O}_X)$ is an isomorphism, therefore $H^1(X,\mathcal{O}_X) \longrightarrow H^1(X,L)$ is surjective and $h^1(L) = 0$ or 1. The same exact sequence, for any D, also yields

$$h^1(L) = -1 + h^0(\mathcal{O}_D) + \dim \mathrm{Im}\{H^1(X,\mathcal{O}_X) \xrightarrow{s} H^1(X,L)\}$$

and a), b) follow at once.

Lemma 2. If $(L^2) > 0$, $h^1(L) = 0$ and $|L|$ contains an irreducible curve C then $h^1(L^{\otimes 2}) = 0$ and $\dim|L^{\otimes 2}| = 2(L^2)$.

Proof. Let $C = \mathrm{div}(s) \in |L|$ be irreducible. Clearly $H^1(X,\mathcal{O}_X) \xrightarrow{s^2} H^1(X,L^{\otimes 2})$ is the 0-map, since it factors through multiplication by s alone and $H^1(X,L) = 0$, hence $s^2 \in \Gamma(L^{\otimes 2})$. On the other hand, C is irreducible and $C^2 > 0$, thus $2C$ is numerically connected and $h^0(\mathcal{O}_{2C}) = 1$. Now Lemma 2 follows from Lemma 1.

In the proof of our proposition, by Lemma 1 we may assume

$$h^1(L) = h^1(L^{\otimes 2}) = 0,$$

$$(L^2) = 2n > 0,\ \dim|L| = n,\ \dim|L^{\otimes 2}| = 4n .$$

Let $\{a_{ij}\} \in Z^1(\mathcal{O}_X)$ be a non-trivial 1-cocycle. We have already seen that the Frobenius cohomology operation yields a relation

$$a_{ij}^2 - \lambda a_{ij} = b_i - b_j$$

with $\lambda \in k$; we shall fix once for all such a datum $\{b_i\}$. Now let $\{f_{ij}\} \in Z^1(\mathcal{O}_X^*)$ be a 1-cocycle representing the class of L in $\mathrm{Pic}(X)$. If $\{s_i\}$ is a section of L, the image of the class of $\{a_{ij}\}$ by

$H^1(X,\mathcal{O}_X) \xrightarrow{s} H^1(X,L)$ is represented by the 1-cocycle $\{s_i a_{ij}\}$ and since $h^1(L) = 0$ this is a 1-coboundary:

$$s_i \, a_{ij} = \sigma_i - f_{ij} \, \sigma_j \, .$$

The corresponding datum $\{\sigma_i\}$ is uniquely determined modulo sections of L, and we fix it by choosing it on a basis of $\Gamma(L)$ and extending by linearity. We use b and σ to construct sections of $L^{\otimes 2}$ as follows: <u>if</u> $s,t \in \Gamma(L)$, <u>if</u> σ,τ <u>are associated data, then</u> $s\tau + \sigma t$ <u>and</u> $\sigma^2 + \lambda\sigma s + bs^2$ <u>are sections of</u> $L^{\otimes 2}$. The verification of this fact is straightforward but clearly uses in an essential way the fact that we work in char. 2.

Now consider sections of the form

$$\text{(A')} \qquad \sigma^2 + \lambda\sigma s + bs^2 + s\tau + \sigma t$$

$$\text{(A'')} \qquad s\tau + \sigma t$$

in $\Gamma(L^{\otimes 2})$. There are some obvious cases in which (A') and (A") vanish identically, namely:

(A'): if $s = \sigma = 0$ or $s = 0, \sigma = t$

(A"): if $s = \mu t, \sigma = \mu\tau$ with $\mu \in k$,

or $s = t = 0$,

or $t = \tau = 0$.

We call these the <u>trivial relations.</u> Now one shows that <u>there exists a non-trivial relation</u>, by a simple dimension argument and the fact that the set of sections we have defined forms a <u>closed subvariety</u> of the vector space $\Gamma(L^{\otimes 2})$. Let (A') = 0 or (A") = 0 be a non-trivial relation; then $\operatorname{div}(s)$ satisfies the conclusion of our proposition. First of all, $s \neq 0$ otherwise $s = \sigma = 0$ and the relation would be trivial. Suppose for example that the relation is

$$\sigma^2 + \lambda\sigma s + bs^2 + s\tau + \sigma t = 0$$

and suppose $\operatorname{div}(s)$ irreducible. We have

$$s(\tau + \lambda\sigma + bs) = \sigma(t + \sigma)$$

and since $\operatorname{div}(s)$ is supposed to be irreducible we easily see that we may deduce that $\sigma_i = s_i g_i$ for suitable regular functions g_i. Now $s_i a_{ij} = \sigma_i - f_{ij}\sigma_j$ yields $s_i a_{ij} = s_i g_i - f_{ij} s_j g_j = s_i(g_i - g_j)$ and $a_{ij} = g_i - g_j$, i.e. $\{a_{ij}\}$ would be a 1-coboundary, contradiction. A similar argument applies for a relation of type (A"). Finally, if $\mathrm{div}(s) = E' + E''$, with $E' \neq E''$ and E', E'' irreducible curves we see that if $s = s's''$ is a corresponding decomposition we may assume that $H^1(X,\mathcal{O}_X) \xrightarrow{s'} H^1(X,\mathcal{O}(E'))$ is the zero map. This however can never happen if E' is non-singular rational, as the cohomology sequence

$$H^0(X,N_{E'}) \longrightarrow H^1(X,\mathcal{O}_X) \xrightarrow{s'} H^1(X,\mathcal{O}(E'))$$

shows, because the normal bundle of E' is negative, hence $H^0(X,N_{E'}) = 0$. This completes the proof.

We shall not dwell further on the study of these interesting surfaces and refer to Bombieri-Mumford for the construction of explicit examples and for additional results.

VI. Characterization of abelian surfaces.

We have encountered in Ch. IV minimal surfaces with $\kappa = 0$, $B_2 = 6$, $B_1 = 4$. We now characterize these surfaces by

Theorem 6. A minimal surface with $\kappa = 0$, $B_2 = 6$, $B_1 = 4$ is abelian.

In order to prove this result we shall make use of still different char. p techniques.

The surface X has invariants $h^1(\mathcal{O}_X) = q = 2$, $p_g = 1$, so that $\mathrm{Pic}^0(X)$ is reduced and Alb(X) is then 2-dimensional. Moreover, $p_g = 1$ shows $\dim |K_X| = 0$, hence either $K_X \equiv 0$ or there is on X a curve of canonical type, namely the curve in $|K_X|$. In this case however X would be elliptic or quasi-elliptic and the canonical bundle formula would show again that $K_X \equiv 0$ or $\kappa = 1$. Hence $K_X \equiv 0$; we deduce that the same holds for every étale covering of X.

Let $f : X \longrightarrow \mathrm{Alb}\, X$ be the Albanese mapping. First of all, if f were

not surjective then f would factorize as $\varphi : X \longrightarrow C$ with C of genus 2. Now C has étale coverings $\pi : C' \longrightarrow C$ of arbitrarily high degree prime to p and $X' = X \times_C C'$ determines an étale covering $X' \longrightarrow X$ together with a surjective map $X' \longrightarrow C'$. Now X' is again a surface of the same type as X while since $X' \longrightarrow C'$ is surjective we certainly have
$q = \dim \mathrm{Alb}(X') \geqq g' = \dim \mathrm{Alb}(C')$. Now $q = 2$ by the Table in Ch. IV, while certainly g' increase to ∞ as the degree of the covering $C' \longrightarrow C$ increases. This shows that $f : X \longrightarrow \mathrm{Alb}\, X$ is surjective. A similar argument now shows that f is a finite map. Suppose E is an irreducible curve with $f(E) = \mathrm{pt}$. Then we must have $(E^2) < 0$ and considering étale coverings $X' \longrightarrow X$ induced by étale coverings of $\mathrm{Alb}\, X$ we obtain X', still in the same class, with arbitrarily many curves E_i with $(E_i^2) < 0$ and the E_i disjoint. This contradicts $B_2(X') = 6$.

Let $\mathrm{Alb}\, X \xrightarrow{\ell} \mathrm{Alb}\, X$ be multiplication by ℓ, let $X_\ell = X \times_{\mathrm{Alb}} \mathrm{Alb}$ and consider

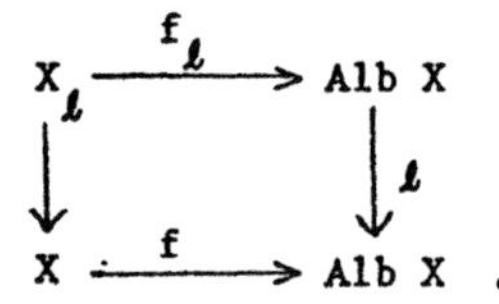

f, and hence f_ℓ, are finite morphisms and $\deg f_\ell = \deg f$. If $(\ell, p) = 1$, which we shall henceforth assume, $X_\ell \longrightarrow X$ and $\mathrm{Alb}\, X \xrightarrow{\ell} \mathrm{Alb}\, X$ are étale maps of degree ℓ^4. If f is a separable map we can argue as follows. Let η be a translation invariant regular 2-form on $\mathrm{Alb}\, X$. Since f is separable, $f^*\eta$ is a regular 2-form on X and $f^*\eta$ never vanishes because $K_X \equiv 0$. It follows at once from this that $f : X \longrightarrow \mathrm{Alb}\, X$ is everywhere an étale morphism; since every étale covering of an abelian variety is an abelian variety, we conclude that X is abelian.

In char. p there is the unpleasant possibility to consider that f may be inseparable; now $f^*\eta$ is identically 0 and the previous argument breaks down. In this case we use the following powerful char. p technique.

The surface X lies in a smooth and proper algebraic family of surfaces defined over a finite field and, if we can prove that the surfaces over the closed points are abelian, X itself will be abelian; this is essentially a specialization argument.

Now suppose that k is a finite field. Fix L ample on Alb X. Then $L_\ell = f_\ell^* L$ is ample on X_ℓ with Hilbert polynomial

$$\chi(L_\ell^{\otimes n}) = (\deg f_\ell)\, \chi(L^{\otimes n})$$
$$= (\deg f)\, \chi(L^{\otimes n})$$

which is independent of ℓ. By a fundamental result of Matsusaka-Mumford there is N, depending only on the Hilbert polynomial, hence independent of ℓ, such that $L_\ell^{\otimes N}$ is very ample for all ℓ. Therefore the infinite set of k-varieties X_ℓ can all be embedded in a fixed $\mathbf{P}^M$, with fixed degree. Now since k is a finite field, there are only finitely many varieties in $\mathbf{P}^M$ with fixed degree, and it follows that all the pairs $(X_\ell, L_\ell^{\otimes N})$ are isomorphic over k to finitely many of them. Now we have:

(i) given X and an ample L, the group of automorphisms f of X with $f^* L \sim L$ is an algebraic group, hence with only finitely many components (Matsusaka);

(ii) the group A_ℓ of translations by points of order ℓ acts on X_ℓ, and each $g \in A_\ell$ has $g^* L_\ell \sim L_\ell$.

Let $(X_n, L_n^{\otimes N})$ be isomorphic to infinitely many $(X_\ell, L_\ell^{\otimes N})$. Then A_ℓ acts on X_n. Let $G_n \subset \mathrm{Aut}(X_n)$ be the group of automorphisms g of X_n such that $g^* L_n \sim L_n$. Then G_n is an algebraic group and $A_\ell \subset G_n$ for infinitely many ℓ. A first consequence is that the connected component G_n^0 of G_n is positive dimensional. Now G_n^0 cannot contain a non-trivial linear subgroup for the action on X_n would make X_n ruled, and $K_{X_n} \equiv 0$. Hence G_n^0 is abelian. Also $A_\ell \cong (\mathbf{Z}/\ell\mathbf{Z})^4$ and subgroups of fixed bounded index of A_ℓ are in G_n^0. We deduce that $\dim G_n^0 \geqq 2$, X_n consists of only one orbit under G_n^0, hence X_n is a coset space G_n^0/H, hence X_n itself is an abelian variety. Finally we have

$$X_n \xrightarrow{\text{étale}} X \xrightarrow{f} \text{Alb } X$$

and X_n, Alb X are both abelian varieties. By choosing an origin, we see that we may assume that the composition $X_n \longrightarrow \text{Alb } X$ is a surjective homomorphism. If K is the kernel, it is now easy to see that $X \cong X_n/K'$ for a suitable subgroup $K' \subset K$, hence X itself is an abelian variety.

VII. Open problems. We mention here rather briefly some interesting open problems of the geometry of surfaces in char. p.

First of all, the Kodaira vanishing theorem in char. p is not yet well understood; further work in this direction is certainly desirable.

In the theory of elliptic and quasi-elliptic surfaces, in char. 2 and 3, one should study how an elliptic fibration may become quasi-elliptic; also one should try to find functional invariants, like the j-invariant in the elliptic case, associated to quasi-elliptic fibrations.

One needs also a deeper study of families of curves in char. p with generic singularities, and a better understanding of unirational surfaces in char. p would be a nice achievement.

The relations (and computation) of the de Rham cohomology and Hodge cohomology in char. p should also be worked out, in order to see how the various degrees of failure of Hodge theory are reflected in the geometry.

From the point of view of Enriques' classification, still much more work has to be done if $\kappa = 1$ and 2. The appearance in these dimensions of the Raynaud surfaces shows that many interesting new aspects may have still to come.

REFERENCES

D. Mumford, Lectures on Curves on an Algebraic Surface, Annals of Math. Studies No. 59, Princeton University Press, 1966.

E. Bombieri and D. Mumford, Enriques' Classification of Surfaces in Char. p, II, in Complex Analysis and Algebraic Geometry, Iwanami Shoten, 1977, pp. 23-42.

——————————, idem, III. Inventiones Math. 35 (1976, pp. 197-232.

D. Mumford, idem, I. in: Global Analysis, Princeton University Press, 1969.

E. Bombieri and D. Husemoller, Classification and embeddings of surfaces, Proc. Symposia Pure Math. XXIX, (1974), pp. 329-420.

CENTRO INTERNAZIONALE MATEMATICO ESTIVO

(C.I.M.E)

To the memory of
Lucien Godeaux /1887-1975/

ALGEBRAIC SURFACES WITH $q = p_g = 0$.

Igor Dolgachev

CONTENTS

Introduction

1. Notations. Let F be a complex algebraic surface. We will use the following standard notations:

$\mathcal{O}_F$: the structure sheaf of F .

$\mathcal{O}_F(D)$: the invertible sheaf associated with a divisor D on F .

$K_F = -c_1(F)$: minus the first Chern class of F or a canonical divisor on F .

$\omega_F = \mathcal{O}_F(K_F)$: the canonical sheaf of F .

$h^i(D)$: the dimension of the space $H^i(F, \mathcal{O}_F(D))$.

$p_g(F) = h^0(K_F) = h^2(\mathcal{O}_F)$; the geometric genus of F .

$q(F) = h^1(K_F) = h^1(\mathcal{O}_F)$: the irregularity of F .

K_F^2 : the self-intersection index of K_F .

$p^{(1)}(F) = K_{F'}^2 + 1$, where F' is a minimal model of a non-rational surface F ; the linear genus of F .

$c_2(F)$: the topological Euler-Poincare characteristic of F .

$p_a(F) = -q(F) + p_g(F) = 1/12(K_F^2 + c_2(F)) - 1$: the arithmetical genus.

$P_n(F) = h^0(nK_F)$: the n-genus of F .

NS(F) : the Neron-Severi group of F, the quotient of the Picard group Pic(F) by the subgroup of divisors algebraically equivalent to zero (= Pic(F) if $q = 0$).

$\mathrm{Tors}(F) = \mathrm{Tors}(NS(F)) = \mathrm{Tors}(H_1(F,\mathbb{Z}))$.

If not stated otherwise F will be always assumed to be non-singular and projective.

2. Historical. It is easily proved that for a rational surface F (that is birationally equivalent to the projective plane $\mathbb{P}^2$) the invariants $q(F)$ and $p_g(F)$ are zero. The interest to non-rational surfaces with vanishing q and p_g was born in 1896 when Castelnuovo had established the necessary and sufficient conditions for a surface to be rational. Clebsh had proved earlier that a curve of genus 0 is rational. The question whether a surface with $q = p_g = 0$ is rational was a natural problem. In [10] Castelnuovo had shown that the answer is negative in general proving that one must add also the condition $P_2 = 0$ and constructing an example of a non-rational surface with $q = p_g = 0$. In the same paper he also exhibited other examples of such surfaces due to Enriques. The latter were of particular destiny, as it turned out later they play a special role in the general classification of algebraic surfaces representing one of the four classes of surfaces with vanishing Kodaira dimension (see [1], [6]) . Both examples of Enriques and Castelnuovo belong to the class of elliptic surfaces, that is they contain a pencil of elliptic curves. In particular, we have for these

surfaces $p^{(1)} = 1$. Later Enriques gave another construction of his surfaces and also presented other non-rational surfaces with $q = p_g = 0$ [17] . They were also elliptic surfaces.

The first examples of surfaces of general type with $q = p_g = 0$ appeared only in 1931-32 when Godeaux had constructed a surface with $q = p_g = 0$ and $p^{(1)} = 2$ [18] and Campedelli had constructed ([9]) a surface with $p^{(1)} = 3$. Later Godeaux constructed some other examples with $p^{(1)} = 3$ [20].

3. Modern development. The new interest to the surfaces under the title is related to the general problem of the existence of surfaces with given topological invariants which became of the main concern after the period of the reconstruction of Enriques' classification results had happily ended. The particular interest to the surfaces with $p^{(1)} = 2$ and 3 (numerical Godeaux and Campedelli surfaces) is due to Bombieri's paper [4] where for all other surfaces it was settled the question of the birationality of the 3-canonical map Φ_{3K} . Now due to works of Bombieri-Catanese [5,II], Miyaoka [32] and Victor Kulikov (non-published) we know that Φ_{3K} is birational for these surfaces, but I do not include the corresponding proofs in this survey refering to the paper of Catanese in these proceedings.

In Chapter II, I expose in more details the results of my paper [14] which deals with elliptic surfaces with $q = p_g = 0$. The theory of Kodaira-Ogg-Šafarevič allows to classify all such surfaces.

In Chapter III, we study more interesting case of surfaces of the general type. All such surfaces are divided into nine classes corresponding to the possible values of $p^{(1)} = 2, 3, \ldots, 10$. To distinguish the surfaces with the same $p^{(1)}$ one may consider the group Tors(F) or more generally the whole fundamental group $\pi_1(F)$. It can be shown (see Chapter III, §6) that there are only a finite number of possible π_1's for surfaces of the same class, and hence one may ask about some explicit estimate of the order of Tors(F) . Unfortunately, this is known only for the cases $p^{(1)} = 2$ (Bombieri) and 3 (Beauville, Reid) and only in the first case this estimate is the best possible. Moreover, we do not know whether the classes with $p^{(1)} = 8$ and 10 are empty* . The examples of surfaces with $4 \leq p^{(1)} \leq 7$ are due to Burniat [7,8] . We present here a new version of his construction ([7], [37]) which enables us to calculate Tors(F) for such surfaces. The examples of surfaces with $p^{(1)} = 9$ are due to Kuga [29] and Beauville [3] .

4. Acknowledgements. This work owes very much to many people with whom I had a conversation on the subject at different periods of my life. It would be impossible to mention them all. I am especially indepted to Miles Reid and Fabrizio Catanese whose critical remarks were very valuable. It is also a great pleasure to thank C.I.M.E. and M.I.T. for their support during the preparation of this paper.

---

* see. Epilogue.

CHAPTER I. CLASSICAL EXAMPLES.

§1. The Enriques surface.

Let $\mathbb{P}^3$ be the projective 3-space with homogeneous coordinates x_i, $i = 0,\dots,3$. Consider the coordinate tetrahedron T : $x_0x_1x_2x_3 = 0$ and let X be a surface in $\mathbb{P}^3$ which passes twicely through the edges E_i $(i = 1,\dots, 6)$ of T, that is, has E_i as its ordinary double lines. We also assume that X has no other singular points outside T and other common points with T. Since the section of F by a coordinate plane is the double reducible cubic curve, we see that F must be of order 6. More explicitly we may consider F as given by the equation:

$$(x_0x_1x_2)^2 + (x_0x_1x_3)^2 + (x_0x_2x_3)^2 + (x_1x_2x_3)^2 + x_0x_1x_2x_3(x_0^2+x_1^2+x_2^2+x_3^2) = 0$$

Let F be the normalization of X. Then F is a non-singular surface. To see it one has to look locally at the normalization of the affine coordinate cross : $xyz = 0$ in $\mathbb{A}^3$. Here the normalization will be just the disjoint union of three planes, the inverse image of the singular loci will be the union of six lines lying by pairs in these planes. Two lines in each of the planes correspond to the two axis lying in the same coordinate plane. The inverse image of the origin will be the three points, each of them is the intersection point of the two lines in one of the planes. So, locally the picture is as follows:

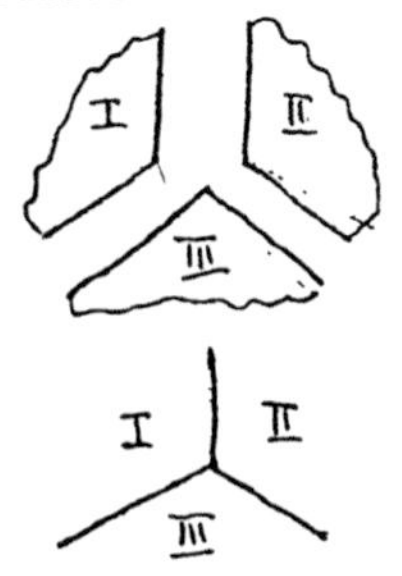

Let $p : F \to X$ be the projection. Then the local analysis above shows that for any edge E_i of the tetrahedron T we have

$$p^{-1}(E_i) = C_i = C_i' + C_i''$$

where C_i' and C_i'' are non-singular rational curves meeting each other transversally at two points arising from the two pinch-points of X lying on each of the edges.

C_i and C_j do not meet if E_i and E_j are not incident, otherwise C_i and C_j meet transversally at one point,

$$C_i \cap C_j \cap C_k = \emptyset \quad \text{for distinct } i, j, k \ .$$

Now we use the classical formula for the canonical sheaf of the normalization of a surface of degree n in $\mathbb{P}^3$:

$$\omega_F = \mathcal{O}_F((n-4)H-\Delta)) \ ,$$

where H is the inverse image of a plane section of X and Δ is the conductor divisor (= the annulator of the sheaf $p_*(\mathcal{O}_F/\mathcal{O}_X)$ (see Mumford's appendix to Chapter III of [43]) . In our case we easily find that

$$\omega_F = \mathcal{O}_F(2H - C) \ ,$$

where $C = C_1 + \ldots + C_6$.

The global sections of ω_F correspond to quadrics in $\mathbb{P}^3$ passing through the edges of the tetrahedron T. Since by trivial reasons such quadrics do not exist we have

$$p_g(F) = h^0(2H - C) = 0 .$$

Next, taking for $2H$ the inverse image of the union of two faces of the tetrahedron, we obtain that

$$K_F \sim 2H - C \sim C + C_i - C_j - C \sim C_i - C_j ,$$

where C_i is the common edge of these faces, and C_j is the opposite edge.

Taking for $4H$ the inverse image of the union of all faces (= the tetrahedron T) we get

$$2K_F \sim 4H - 2C \sim 2C_i - 2C_j \sim 0 .$$

Thus we have

$$P_2(F) = h^0(\mathcal{O}_F) = 1$$

and hence F is non-rational.

Since K_F is numerically equivalent to zero, we have

$$C_i^2 = C_i K_F = C_i(C_i - C_j) = 0 , \quad i + 1,\ldots, 6 .$$

By the adjunction formula we get

$$H^0(C_j, \mathcal{O}_{C_j}) = H^0(C_j, \mathcal{O}_{C_j}) = \mathbf{C} .$$

Thus, C_i is a reducible curve of arithmetical genus 1 . Since $2C_i \sim 2C_j$ and C_i does not meet C_j we infer that the linear system $|2C_i|$ contains a pencil of curves of arithmetical genus 1 Since there are no base points of $2C_i$ we obtain by Bertini's theorem that almost all curves form this pencil are non-singular elliptic curves. Note also that this pencil contains two degenerate curves, $2C_i$ and $2C_j$.

Now we may use the formula expressing $c_2(F)$ in terms of the Euler-Poincare characteristic of degenerate curves of the elliptic pencil (see [1], Ch. IV):

$$c_2(F) = \sum_i \chi(B_i)$$

where B_i are all singular curves of the pencil. Since

$$\chi(2C_i) = \chi(2C_j) = \chi(C_i) = 1$$

we deduce that

$$c_2(F) > 0 .$$

Since $K_F^2 = 0$ we get by the Noether formula $12(I - q(F)) = c_2(F) > 0$. This obviously implies that $q(F) = 0$.

§2. The Godeaux surface.

Consider the projective involution σ of $\mathbb{P}^3$ of order 5 given in coordinates by the formula:

$$(x_0,x_1,x_2,x_3) \longmapsto (x_0,\zeta x_1,\zeta^2 x_2,\zeta^3 x_3) ,$$

ζ being a primitive 5-th root of unity. This involution acts freely outside the vertices of the coordinate tetrahedron. Let F' be a non-singular quintic which is invariant under σ and does not pass through these vertices. For example, we may take for F' a quintic with the equation:

$$a_0x_0^5 + a_1x_1^5 + a_2x_2^5 + a_3x_3^5 = 0 .$$

(For a general surface F' with the properties above one has to add to the left side 8 invariant monomials $x_0x_2^2x_3^2,\ldots$). Let G be the cyclic group of order 5 generated by σ, acting freely on F'. Consider the quotient $F = F'/G$, the projection $p : F' \to F$ is a finite non-ramified map of non-singular surfaces.

<u>Lemma</u>. Let $p : F' \to F$ be a finite non-ramified map of degree n. Then

$$K_{F'}^2 = n\, K_F^2$$

$$1 + p_a(F') = n\,(1 + p_a(F)) .$$

<u>Proof</u>. The first relation easily follows from the equality $p^*(\omega_F) = \omega_{F'}$, since p is smooth and finite. The second one follows from the Noether formula and the relation $c_2(F') = n\, c_2(F)$, which can be proved either by topological arguments or using the equality

$$p^*(\Omega^1_F) = \Omega^1_{F'} ,$$

Ω^1 being the sheaf of 1-differentials, and standard properties of Chern classes.

Since we have for F', $K^2_{F'} = 5$, $p_a(F') = 4$ we get from the lemma

$$K^2_F = 1 , \quad p_a(F) = 0 .$$

Since, obviously, $q(F) \leq q(F')$, we obtain

$$p_g(F) = 0 .$$

Next, note that F is minimal, that is there are no exceptional curves of the first kind lying on it. Indeed, the inverse image of such curve under p would be the disjoint union of five exceptional curves of the first kind on F'. However, F' is minimal. From the minimality of F and the fact $K^2_F \geq 1$ it follows that F is of general type. Another way to show this is to use the property of ample sheaves: $p^*(\omega_F)$ is ample implies ω_F is ample.

Since F' is simply-connected we obtain that the map p is the universal covering. In particular, $\mathrm{Tors}(F) = \pi_1(F) = \mathbb{Z}/5\mathbb{Z}$.

§3. The Campedelli surface.

This is a double ramified covering of the projective plane $\mathbb{P}^2$ branched along some curve of the IO-th degree (more precisely it is a minimal non-singular model of such covering).

Let W be the following reducible curve of the IO-th degree

$$W = C_1 \cup C_2 \cup C_3 \cup D \quad ,$$

where C_i are non-singular conics and D is a non-singular quartic with the following properties:

$$C_1 \cap C_2 = 2P_1 + 2P_2 ; \quad C_1 \cap C_3 = 2P_3 + 2P_4 ; \quad C_2 \cap C_3 = 2P_5 + 2P_6$$

$$D \cap C_1 = 2(P_1 + P_2 + P_3 + P_4) ; \quad D \cap C_2 = 2(P_1 + P_2 + P_5 + P_6)$$

$$D \cap C_3 = 2(P_3 + P_4 + P_5 + P_6) .$$

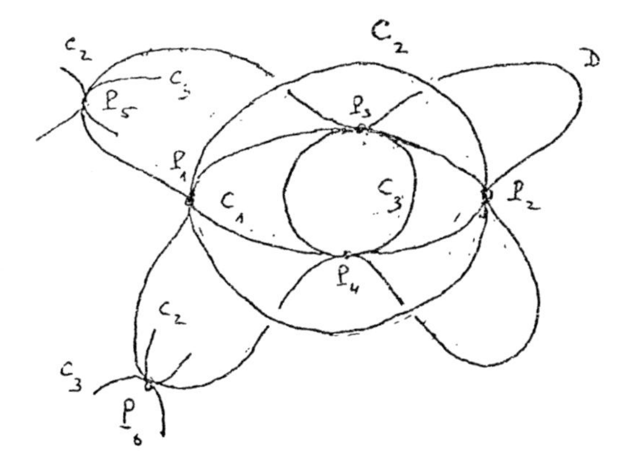

To see that such configuration of curves exists one may take for C_2 and C_3 two concentric circles lying in the complement to the line at infinity, the points P_5 and P_6 will be the two cyclic points. The existence of a quartic D touching the conics C_i easily follows from the consideration of the net $\lambda C_1C_2 + \mu C_1C_3 + \nu C_2C_3 = 0$.

Lemma 1. Let X be a non-singular surface and W a reduced curve on it. Suppose that there exists a divisor D on X such that $W \sim 2D$, then there is a double covering

$$f : Y \to X$$

branched exactly along W . Moreover, Y is normal and non-singular over the complement to the singular focus of W .

Proof. Assume firstly that W is non-singular. Let F be the line bundle corresponding to the divisor D and (U_j) a coordinate covering of X such that $F|U_j$ is trivial and W is given by the local equation $\{c_j = 0\}$ on U_j . Let g_{ij} be a system of transition functions for F , then $c_i = g_{ij}^2 c_j$ on $U_i \cap U_j$ and we may consider the subvariety Y of F given by the equations $x_j^2 = c_j$, where x_j is a fibre coordinate of $F|U_j$. It is obviously checked that the projection $Y \to X$ satisfies the properties stated in the lemma.

If W is singular we apply the arguments above to X replaced by $X' = X - S$ and W by $W' = W - S$, where S is the singular locus of W . Then it suffices to take for Y the normalization of X in the double covering $Y' \to X'$ constructed as above.

Remark. The sheaf $L = \mathcal{O}_X(D)$ can be characterized as the subsheaf of antiinvariant sections of the direct image $f_*(\mathcal{O}_Y)$. If $q(X) = 0$ then this sheaf is determined uniquely by W (since they differ by an element of order 2 in $\mathrm{Pic}(X)$) . This shows that in this case any double covering with properties from lemma 1 can be obtained by the construction of the lemma.

Applying this lemma to the plane $\mathbb{P}^2$ and the 10-th degree curve W we may construct a double covering Y of $\mathbb{P}^2$ branched along W . This surface has six singular points lying over the points P_i . The Campedelli surface F will be obtained as the minimal non-singular model of Y .

Let $p : X \to \mathbb{P}^2$ be the minimal resolution of singularities of the curve W . The proper transform of W is given by

$$p^{-1}(W) \sim p^*(10L) - 3\sum_{i=1}^{6} S_i - 6\sum_{i=1}^{6} S_i' ,$$

where L is a line on $\mathbb{P}^2$,

$$p^{-1}(P_i) = S_i + S_i' , \quad i = 1,\dots,6$$

with $S_i^2 = -2$, $S_i'^2 = -1$.

Now we apply the lemma to the surface X and the non-singular curve

$$p^{-1}(W) + \sum_{i=1}^{6} S_i$$

and consider the corresponding double covering $r : F' \to X$. To compute the canonical class $K_{F'}$ we use the following:

Lemma 2. Let $g : V' \to V$ be the double covering of non-singular surfaces branched along the curve W , $g^*(W) \sim 2\bar{D}$ for some divisor $\bar{D}$ on V' . Then

$$K_{V'} \sim g^*(K_V) + \bar{D} .$$

Proof. First, note that our double covering can be obtained by the construction from lemma 1 . In fact, consider the splitting

$$g_*(\mathcal{O}_{V'}) = \mathcal{O}_V \oplus L$$

into invariant and anti-invariant pieces. Then clearly $L^{\otimes 2}$ is contained in the invariant piece that is in $\mathcal{O}_V$. Thus $\mathcal{O}_V$-Algebra $g_*(\mathcal{O}_{V'})$ is the quotient algebra of the symmetric algebra $\mathrm{Symm}(L) = \mathcal{O}_V \oplus L \oplus L^{\otimes 2} \oplus \dots$ by the Ideal generated by $L^{\otimes 2} - J$, where J is an ideal sheaf in $\mathcal{O}_V$. Taking the spectrums we get that $V' = \mathrm{Spec}(g_*(\mathcal{O}_{V'}))$ is isomorphic to the closed subscheme of the line bundle $F = \mathbf{V}(\check{L}) = \mathrm{Spec}(\mathrm{Symm}(L))$. Looking locally we easily identify J with the sheaf $\mathcal{O}_V(-W)$ and obtain that V' is constructed with the help of a divisor D corresponding to F in the same way as in lemma 1 .

Now, the formula for $K_{V'}$ can be proved very simply. In notations of lemma 1 we consider a 2-form w on V in local coordinates c_j and some other function t_j . Then we use the relation $dc_j \wedge dt_j = 2x_j dx_j \wedge dt_j$ to obtain that $(g^*(W)) = g^*((W)) +$ $+ \{x_j = 0\}$ (the brackets ( ) denotes the divisor of a 2-form). This proves the lemma.

Thus, we have

$$K_{F'} \sim r^*(-3p^*(L) + \sum_{i=1}^{6} S_i + 2\sum_{i=1}^{6} S_i') + \frac{1}{2} r^*(10p^*(L) - 2\sum_{i=1}^{6} S_i - 6\sum_{i=1}^{6} S_i'$$

$$\sim r^*(2p^*(L) - \sum_{i=1}^{6} S_i') .$$

Assume that $D \in |K_{F'}|$, then we see from above that $D = r^*(D')$ for some

$$D' \in |2p^*(L) - \sum_{i=1}^{6} S_i'| .$$

The latter linear system is equal to the inverse transform of the system of conics on $\mathbb{P}^2$ passing through all points P_i. To show that the latter does not exist we argue as follows. Taking for P_1 and P_2 the two cyclic infinite points $(1, \pm i, 0)$ we get the equations for C_1 and C_2 in the form:

$$X_1^2 + X_2^2 - a^2X_0^2 = 0 , \quad X_1^2 + X_2^2 - b^2X_0^2 = 0$$

and the equation for C_3 on the form:

$$\frac{X_1^2}{a^2} + \frac{X_2^2}{b^2} - X_0^2 = 0 .$$

The points P_3, P_4, P_5, P_6 will have the coordinates $(\pm a, 0, 1)$, $(0, \pm b, 1)$ respectively. Now let C be a conic with an equation

$$a_1x_1^2 + a_2x_2^2 + a_3x_1x_2 + a_4x_1x_0 + a_5X_2X_0 + a_6x_0^2 = 0$$

which passes through the points $P_1,\ldots,P_6$. Since it passes through P_1 and P_2 we may assume that $a_3 = 0$ and $a_1 = a_2 = 1$. Since it passes through P_3 and P_4 we get the equations

$$a^2 \pm a_4 + a_6 = 0$$

which give $a_4 = 0$ and $a_6 = -a^2$. Similarly we get $a_5 = 0$ and $a_6 = -b^2$. This contradiction shows that C does not exist.

Thus $K_{F'} = \emptyset$ and $P_g'(F') = 0$.

Since r is branched along S_i, $i = 1,\ldots, 6$ and $p^{-1}(C_i)$, $i = 1, 2, 3$, we see that

$$r^*(S_i) = 2\,\bar{S}_i \ , \quad r^*(p^{-1}(C_i)) = 2\,\bar{C}_i$$

for some curves $\bar{S}_i$ and $\bar{C}_i'$ on F' . Also, we notice that

$$\bar{S}_i^2 = \frac{1}{4}(r^*(S_i)^2) = \frac{1}{4}(2S_i^2) = \frac{1}{4}(-4) = -1 \ ,$$

$$\bar{C}_i'^2 = \frac{1}{4}(r^*(p^{-1}(C_i))^2) = \frac{1}{4}(2(p^{-1}(C_i))^2) = \frac{1}{4}(-8) = -2 \ .$$

This shows that $\overline{S}_i$ are exceptional curves of the 1st kind. Let $\sigma : F' \to F$ be the blowing down of all $\overline{S}_i$. We will show that F is a minimal model of F' . We have

$$2K_{F'} \sim r^{*}(4p^{*}(L) - 2\sum_{i=1}^{6} S_i') \sim$$

$$\sim 2\overline{C}_1' + 2\overline{C}_2' + 2S_1' + 2S_2' + 2(\sum_{i\neq 5,6} \overline{S}_i + \overline{S}_1 + \overline{S}_2)$$

$$\sim 2\overline{C}_1' + 2\overline{C}_3' + 2S_3' + 2S_4' + 2(\sum_{i\neq 3,4} \overline{S}_i + \overline{S}_3 + \overline{S}_4)$$

$$\sim 2\overline{C}_2' + 2\overline{C}_3' + 2S_5' + 2S_6' + 2(\sum_{i\neq 1,2} \overline{S}_i + \overline{S}_5 + \overline{S}_6)$$

since

$$p^{-1}(C_1) \sim p^{*}(2L) - 2\sum_{i\neq 5,6} S_i' - \sum_{i\neq 5,6} S_i$$

$$p^{-1}(C_2) \sim p^{*}(2L) - 2\sum_{i\neq 3,4} S_i' - \sum_{i\neq 3,4} S_i$$

$$p^{-1}(C_3) \sim p^{*}(2L) - 2\sum_{i=1,2} S_i' - \sum_{i\neq 1,2} S_i'$$

This shows that

$$2K_F \sim 2\overline{C}_1 + 2\overline{C}_2 + 2\overline{S}_1 + 2\overline{S}_2 \sim$$

$$\sim 2\overline{C}_2 + 2\overline{C}_3 + 2\overline{S}_3 + 2\overline{S}_4 \sim$$

$$\sim 2\overline{C}_1 + 2\overline{C}_3 + 2\overline{S}_5 + 2\overline{S}_6 \ ,$$

where $\overline{C}_i = \sigma_{*}(\overline{C}_i')$, $\overline{S}_i = \sigma_{*}(S_i')$.

If E is an exceptional curve of the 1st kind on F , then $(E\,K_F) = -1$ and hence E must coincide with one of the curves $\overline{C}_i$ or $\overline{S}_i$. However, neither of them is an exceptional curve, because

$$\overline{C}_i^2 = \overline{C}_i'^2 = -2 \ , \quad (\overline{S}_i K_F) = 1 \quad .$$

To compute K_F^2 we use that

$$K_F^2 = (\overline{C}_1 + \overline{C}_2 + \overline{S}_1 + \overline{S}_2)^2 = -2 - 2 - 1 - 1 + 8 = 2 \ .$$

It remains to notice that F is a surface of general type, since it is minimal and has positive K_F^2 . In particular, we have $q(F) \leq p_g(F) = 0$ (see Chap. 3, §1, lemma 3) . Also note that $2K_F$ is determined by the net of quartics $\lambda C_1C_2 + \mu C_1C_3 + \nu C_2C_3$ and is of dimension 2 .

We also have the following obvious torsion divisors of order 2 on F :

$$g_1 : K_F - \overline{C}_1 - \overline{C}_2 - \overline{S}_1 - \overline{S}_2 \ ,$$

$$g_2 : K_F - \overline{C}_1 - \overline{C}_3 - \overline{S}_5 - \overline{S}_6 \ ,$$

$$g_3 : K_F - \overline{C}_2 - \overline{C}_3 - \overline{S}_3 - \overline{S}_4 \ ,$$

$$g_4 : K_F - \overline{D} \ , \text{ where } \sigma_*(r^*(p^{-1}(D))) = 2\overline{D} \quad .$$

It is immediately checked that

$$g_1 + g_2 + g_3 = 0$$

and

$$g_1 + g_2 + g_4 = g_3 + g_4 \not\sim 0 .$$

This shows that

$$\mathrm{Tors}(F) \supseteq (Z/2Z)^3 .$$

It will be shown in Chapter III, §3 that, in fact, we have the equality

$$\mathrm{Tors}(F) = (Z/2Z)^3 .$$

CHAPTER 2. ELLIPTIC SURFACES.

1. Generalities.

A projective non-singulur surface X is called elliptic if there exists a morphism $f : X \to B$ onto a non-singular curve B whose general fibre X is a smooth curve of genus 1. Such f is called an elliptic fibration on X. From general properties of morphisms of schemes we infer that almost all fibres are non-singular elliptic curves over the ground field k (as everywhere in this paper we assume that $k = \mathbb{C}$ or algebraically closed of characteristic 0). An elliptic surface X is called minimal if there exist an elliptic fibration without exceptional curves of the 1st kind in its fibres ((such fibration will be called minimal).

Let $f : X \to B$ be an elliptic fibration on an elliptic surface X and X_b a fibre over a point $b \in B$. Consider X_b as a positive divisor on X, then according to Kodaira [26] it is one of the following types:

$m^1 0$: $X_b = mE_0$, $m \geq 1$, where E_0 is a non-singular elliptic curve;

$m^1 1$: $X_b = mE_0$, $m \geq 1$, where E_0 is a rational curve with a node;

$m^1 2$: $X_b = mE_0 + mE_1$, $m \geq 1$, where E_0 and E_1 are non-singular rational curves meeting transversally at two points;

$m^1 b$: $X_b = mE_0 + \ldots + mE_{b-1}$, $m \geq 1$, where E_i are rational non-singular curves with $E_i \cap E_j \cap E_k = \emptyset$ for distinct i, j, and k and $(E_iE_{i+1}) = 1$, $i = 0, \ldots, b-1$, assuming $E_b = E_0$ $(b \geq 3)$.

II : $X_b = E_0$, a rational curve with a cusp;

III : $X_b = E_0 + E_1$, where E_0 and E_1 are non-singular rational curves with simple contact at one point;

IV : $X_b = E_0 + E_1 + E_2$, where E_i are non-singular rational curves transversally meeting each other at one point $p = E_0 \cap E_1 \cap E_2$;

I_b^* : $X_b = E_0 + E_1 + E_2 + E_3 + 2E_{4+b}$, where all E_i are non-singular rational curves transversally interesecting as shown on the picture

II^* : $X_b = E_0 + 2E_1 + 3E_2 + 4E_3 + 5E_4 + 2E_5 + 4E_6 + 3E_7 + 6E_8$, where E_i are non-singular rational curves interesecting as shown on the picture

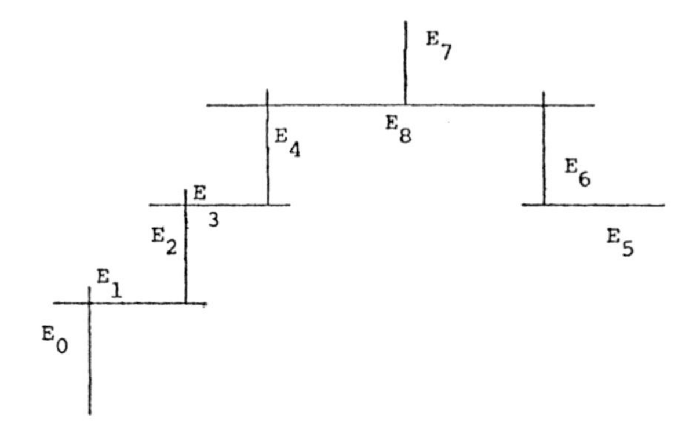

III^* : $X_b = E_0 + 2E_1 + 3E_2 + E_3 + 2E_4 + 3E_5 + 2E_6 + 4E_7$, where E_i are non-singular rational curves and the picture is

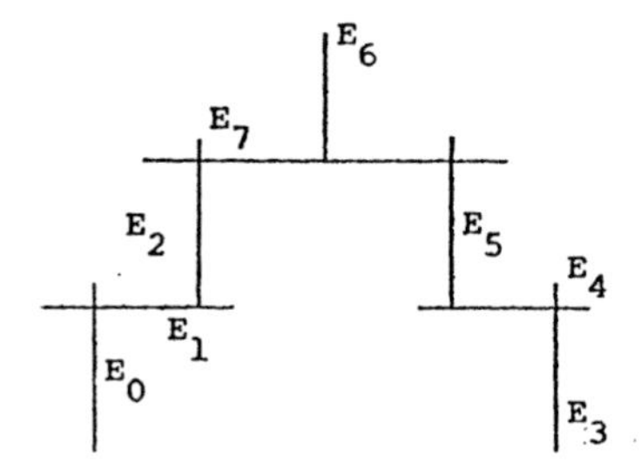

IV^* : $X_b = E_0 + 2E_1 + E_2 + 2E_3 + E_4 + 2E_5 + 3E_6$, where E_i are rational non-singular curves and the picture is

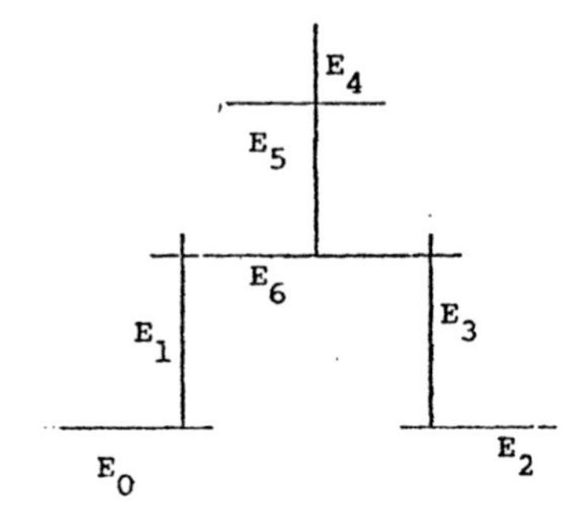

A singular fibre of type m^1b , $b \geq 0$, $m \geq 2$, is called <u>multiple</u> of multiplicity m .

Let $f : X \to B$ be an elliptic fibration, then its general fibre X_η is a sooth curve of genus 1 over the field K of rational functions of B and it is an abelian variety over K if and only if it has a K-rational point. In geometric terms the latter is equivalent to the existence of a global section $s : B \to X$ of the morphism f .

Consider the jacobian variety J of X_η , this is again a smooth curve over K of genus 1 with a rational point over K . For any extension K'/K such that X has a K'-rational point there exists a natural isomorphism of K'-curves $X_\eta \otimes_K K' \simeq J \otimes_K K'$. According to general properties of birational transformations of two-dimensional schemes there exists the unique minimal elliptic fibration $j : A \to B$ such that $A_\eta \simeq J$. This surface is called the jacobian surface of the elliptic surface X . Since j has a section, all singular fibres of j are non-multiple.

Proposition 1. For any $b \in B$ such that X_b is a non-multiple fibre the fibrations $f : X \to B$ and $j : A \to B$ are isomorphic over the henselization $\tilde{O}_b$ of the local ring $O_{B,b}$.

Proof. Let $\tilde{f}_b : \tilde{X}(b) \to \mathrm{Spec}\, \tilde{O}_b$ be the restriction of f over $\tilde{O}_b$, and $\tilde{j}_b : \tilde{A}(b) \to \mathrm{Spec}\, \tilde{O}_b$ the same for j . Since $\tilde{f}_b$ is smooth at some point of a component of multiplicity 1, there exists a section of $\tilde{f}_b$. This implies that the general fibre $\tilde{X}(b)_\eta$ is an abelian curve over the fraction field $\tilde{K}_b$ of $\tilde{O}_b$. From this we infer easily that $\tilde{X}(b)_\eta \simeq \tilde{A}(b)_\eta$ and hence in virtue of the uniqueness of the minimal models we get $\tilde{X}(b) \simeq \tilde{A}(b)$.

Proposition 2. Let $b \in B$ such that X_b is a multiple fibre of type ${}_m1_b$. Then the fibre A_b of $j : A \to B$ is of type ${}_1A_b$.

Proof. Let $B' \to B$ be a covering of B ramified at some point $b' \in B'$ over b with the ramification index equal to m .

Let $f'_{b'} : X'(b') \to \operatorname{Spec} O_{B',b'}$ be the restriction of the base change map $X \times_B B' \to B'$ over the local ring $O_{B',b'}$. Denote by $\overline{X}'(b')$ the normalization of $X'(b')$ and let $\overline{f}'_{b'} : \overline{X}'(b') \to \operatorname{Spec} O_{B',b'}$ be the composite map.

Let $x \in X$ be a double point of the fibre X_b. Then formally at x the map $f : X \to B$ is isomorphic to the map $\mathbb{A}^2 \to \mathbb{A}^1$ given by $t = (xy)^m$. This shows that $X'(b')$ formally at the point x' lying over x is isomorphic to the hypersurface $t^m = (xy)^m$ in $\mathbb{A}^3$. Taking the normalization we observe that there are exactly m points $x'_1, \ldots, x'_m \in \overline{X}'(b')$ lying over x' and formally at each x'_i $\overline{X}'(b')$ is given by the equation $t = xy$. Looking globally we infer that the fibre $\overline{X}(b')_0$ is of type 1^1mb.

Performing the same base change for $j : A \to B$ and resolving the singularities of the obtained surface $A'(b')$ we will get the scheme over $O_{B',b'}$ with the closed fibre of type 1^1mb (Proposition 1). Checking case by case we find that it can be only if the fibre of j over b is of type 1^1b.

Let $j : A \to B$ be a minimal elliptic fibration with a global section, $W(j)$ be the set of all minimal elliptic fibrations over B for which j serves as the jacobian fibration. For any $f : X \to B$ from $W(j)$ the general fibre X_K is a principal homogeneous space (p.h.s) for A_K over the field K of rational functions on B. As it is well known the set of all p.h.s. for A_K forms the Galois cohomology group $H^1(K, A_K)$. In virtue of the existence and uniqueness of minimal models for A_K the map $W(j) \to H^1(K, A_K)$ is bijective.

To compute $H^1(K, A_K)$ we argue as follows ([34, 38, 41]). Let $i : \eta = \operatorname{Spec} K \hookrightarrow B$ be the inclusion of the general point. Identify A_K with the etale sheaf which it represents and let $\underline{A} = i_* A_K$. The sheaf $\underline{A}$ is representable by the commutative group scheme over B which is obtained by throwing out all points of the surface A where f is non-smooth (the <u>Neron model</u> of A_K). The Leray spectral sequence for i gives the exact cohomology sequence:

$$0 \longrightarrow H^1(B,\underline{A}) \longrightarrow H^1(K,A_K) \xrightarrow{\psi} H^0(B,R^1 i_* A_K) \longrightarrow H^2(B,\underline{A})$$

For any closed point $b \in B$ we have

$$(R^1 i_* A_K)_b = H^1(\tilde{K}_b, A_{\tilde{K}_b}) \quad ,$$

where $\tilde{K}_b$ is the fraction field of the henselization $\tilde{O}_{B,b}$ of the local ring $O_{B,b}$, $A_{\tilde{K}_b} = A_K \times_K \tilde{K}_b$. To compute $H^1(\tilde{K}_b, A_{\tilde{K}_b})$ it suffices to compute for all n the subgroup ${}_nH^1(\tilde{K}_b, A_{\tilde{K}_b})$ of the elements killed by multiplication by n (since H^1 is always periodical). Using the Kummer exact sequence

$$0 \longrightarrow {}_nA_{\tilde{K}_b} \longrightarrow A_{\tilde{K}_b} \xrightarrow{n} A_{\tilde{K}_b} \longrightarrow 0$$

we get

$${}_nH^1(\tilde{K}_b, A_{\tilde{K}_b}) = H^1(\tilde{K}_b, {}_nA_{\tilde{K}_b}) \quad .$$

Now, since $A_{\tilde{K}_b}$ coincides with its Picard variety $\mathrm{Pic}^0(A_{\tilde{K}_b})$, we have

$$ {}_nA_{\tilde{K}_b} = ({}_nR^1\tilde{j}^b_* G_m)_{\tilde{K}_b} = (R^1\tilde{j}^b_*\mu_n)_{\tilde{K}_b} $$

where $\tilde{j}^b : \tilde{A}(b) \to \mathrm{Spec}\,\tilde{O}_{B,b}$ is the strict localization of the morphism j over b. Since $\mathrm{Spec}\,\tilde{O}_{B,b}$ is cohomologically trivial

$$ H^1(\tilde{K}_b, {}_nA_{\tilde{K}_b}) = H^1(\tilde{K}_b, i_b^* R^1\tilde{j}^b_*\mu_n) = H^0(\mathrm{Spec}\,\tilde{O}_{B,b}, R^1\tilde{j}^b_*\mu_n) = $$

$$ = H^1(\tilde{A}(b), \mu_n) = H^1(A_b, \mu_n), $$

where $i_b : \mathrm{Spec}\tilde{K}_b \hookrightarrow \mathrm{Spec}\tilde{O}_{B,b}$ is the inclusion of the general point. It remains to add that

$$ H^1(A_b, \mu_n) = (\mathbb{Z}/n\mathbb{Z})^{d_b}, $$

where $d_b = 2$ in case A_b is of type ${}_1I_0$, $d_b = 1$ in case A_b is of type ${}_1I_l$, $l \geq 1$, $d_b = 0$ in the remaining cases.

Let $f : X \to B$ be the elliptic fibration representing an element $x \in H^1(K, A_K)$. Then we interprete the composite map

$$ \psi_b : H^1(K, A_K) \longrightarrow H^0(B, R^1 i_* A_K) \longrightarrow (R^1 i_* A_K)_b = H^1(\tilde{K}_b, A_{\tilde{K}_b}) $$

as follows. The general fibre of the strict localization $\tilde{f}^b : \tilde{X}(b) \to \mathrm{Spec}\,\tilde{O}_{B,b}$ represents a p.h.s. for $A_{\tilde{K}_b}$ over the field $\tilde{K}_b$ and, hence, an element of $H^1(\tilde{K}_b, A_{\tilde{K}_b})$. Also, it can be checked that $\psi_b(X)$ equals the class of the normal sheaf of the reduced fibre X_b^0 in the Picard group $\mathrm{Pic}(X_b^0)$, whose torsion part is identified with $H^1(K_b, A_{K_b}) = \varinjlim H^1(A_b, \mu_n) = \varinjlim H^1(X_b^0, \mu_n)$.

From this observation we immediately obtain the following

Proposition 3. For any $x \in H^1(K, A_K)$ $\psi_b(x) \neq 0$ if and only if the fibre X_b of the corresponding elliptic fibration is multiple. The multiplicity of X_b equals the order of $\psi_b(x)$ in $H^1(\tilde{K}_b, A_{\tilde{K}_b})$.

The last assertion follows from the proof of Proposition 2.

Now we shall compute the kernel $H^1(B,\underline{A})$ of the map (so called the Tate-Shafarevich group of A_K) .. First, we have the following exact sequence:

$$0 \longrightarrow \underline{A} \longrightarrow R^1 j_* \mathbb{G}_m \longrightarrow \mathbb{Z}_B \longrightarrow 0$$

which comes from the identification of A_K with its Picard variety $\underline{\mathrm{Pic}}^0_{A_K} = \mathrm{Ker}((R^1 j_* \mathbb{G}_m)_K \to \mathbb{Z}_K)$ (see the details in [24]) . Since j has a global section, the exact cohomology sequence gives the isomorphism

$$H^1(B,\underline{A}) = H^1(B, R^1 j_* \mathbb{G}_m) .$$

Next, considering the Leray spectral sequence for j and $G_{m,A}$ and using that $R^i j_* G_m = H^1(B, G_m) = 0$, $i > 1$ we get

$$H^1(B,\underline{A}) = \mathrm{Br}(A) = H^2(A, G_m) .$$

In virtue of birational invariance of Br ([24]) we obtain

Proposition 4. Assume that A is a rational surface, then

$$H^1(B,A) = 0 .$$

In particular, any minimal elliptic surface without multiple fibers whose jacobian surface A is rational is isomorphic to A.

The last thing to do is to investigate the group $H^2(B,\underline{A})$. Let $\underline{A}^0$ be the subsheaf of $\underline{A}$ which is representable by the connected component of the unit of the group scheme $\underline{A}$ (equal to the surface A minus all irreducible components of the fibers which do not meet some fixed section of j and also minus singular points of irreducible fibers). We have the "Kummer exact sequence"

$$0 \longrightarrow {}_n\underline{A}^0 \longrightarrow \underline{A}^0 \xrightarrow{n} \underline{A}^0 \longrightarrow 0 ,$$

which gives the exact sequence

$$H^1(B, \underline{A}^0) \longrightarrow H^2(B, {}_n\underline{A}^0) \longrightarrow {}_nH^2(B, \underline{A}^0) \longrightarrow 0 .$$

The quotient sheaf $\underline{A}/\underline{A}^0$ has finite support, hence

$$H^i(B,\underline{A}) = H^i(B,\underline{A}^0) , \quad i \geq 1 .$$

Applying the global duality theorem [12], we get

$$H^2(B,{}_n\underline{A}^0) = \mathrm{Hom}(H^0(B,{}_n\hat{\underline{A}}^0) , Z/nZ) .$$

The dual sheaf ${}_n\hat{A}$ coincides with ${}_nA$ in virtue of the auto-duality of the jacobian variety A_K. Now we use the following.

Lemma. Suppose that $q(A) = 0$. Then

$$H^0(B, {}_nA^0) = 0 .$$

Proof. Any element of the group $H^0(B, {}_nA^0)$ represents a section of j of order dividing n which meets the same irreducible component of a fiber as the fixed zero section. Moreover, any two such sections do not meet each other, since for any point $b \in B$ the reduction homomorphism $A(\tilde{K}_b) \to A_b(k)$ is an isomorphism on the subgroup of points of finite order. The latter follows, for example, from the equality $H^0(\mathrm{Spec}\, \tilde{O}_{B,b}, {}_nA) = H^0(k, {}_nA_b)$ which is a particular case of some general property of étale cohomology ([12]). Suppose that $H^0(B, {}_nA^0) \neq 0$, and let S be a section from this group different from the zero section S_0. Then

$$n(S - S_0) \sim \sum m_i F_i'$$

where F_i' is a divisor supported in some fiber F_i of j. Since S and S_0 meet the same component of fibers we get immediately that $F_i' \theta = 0$ for each component θ of F_i. Applying the main lemma below we get that $F_i' = F_i$ and hence

$$n(S - S_0) \sim kF$$

where F is any fiber (use that since $q(A) = 0$ we have $B = \mathbb{P}^1$ and hence all fibers are linearly equivalent). Now from the computation of K_A (see again below) we get for any section $(S \cdot K_A) = -1 + p_g(A) < 0$. Since $S \simeq \mathbb{P}^1$ we get $S^2 = -2 + 1 - p_g(A) = -1 - p_g(A) < 0$.. However

$$(n(S - S_0))^2 = n^2S^2 + n^2S_0^2 = k^2(F^2) = 0 .$$

This contradiction proves the lemma.

From this lemma we get the following.

Proposition 5. Suppose that $q(A) = 0$. Then the map

$$\psi : H^1(K, A_K)' \longrightarrow \bigoplus_{b \in B} H^1(K_b, A_{K_b})$$

is surjective. In particular, for any finite set of closed points $b_1, \ldots, b_r \in B$ such that the fiber A_{b_i} is of type ${}_1I_{h_i}$ $(i=1,\ldots r;$ $h_i > 0)$ and any collection of positive numbers $m_1, \ldots, m_r$ there exists a minimal elliptic fibration $f : X \to B$ whose jacobian fibration equals j and whose fibers X_{b_i} are of type ${}_{m_i}I_{h_i}$, $i=1,\ldots,r$.

Now we shall compute the canonical class K_X of an elliptic surface X . We restrict ourselves for the simplicity to the case of regular surfaces X (i.e. we assume that $q(X) = 0$) . For the general case we refer to [6] or [27] . In particular, we may assume that the base B of any elliptic fibration $f : X \to B$ is the projective line $\mathbb{P}^1$.

Main lemma. ([6]). Let $C = \sum n_i C_i$ be an effective divisor on a surface X with each E_i irreducible. Assume that

$$(C_i, D) \leqq 0 \text{ , all } i$$

and that D is connected.

Then every divisor $Z = \sum m_i C_i$ satisfies $Z^2 \leqq 0$ and equality holds if and only if $D^2 = 0$ and $Z = rD$, $r \in \mathbb{Q}$.

Proof. Write $x_i = m_i/n_i$ and consider the equality

$$\begin{aligned} Z^2 &= \sum x_i x_j n_i n_j (C_i \cdot C_j) \\ &\leqq \sum x_i^2 n_i^2 (C_i C_i) + \sum_{i \neq j} \frac{1}{2}(x_i^2 + x_j^2) n_i n_j (C_i C_j) \\ &= \sum x_i^2 n_i (C_i D) \leqq 0 \text{ .} \end{aligned}$$

If equality holds everywhere, then we have either $x_i = x_j$ or $(C_i C_j) = 0$ for all $i;j$; since D is connected the last possiblity does not occur. Hence x_i is constant, that means that $m_i = rn_i$, $r \in \mathbb{Q}$.

Theorem. Let $f : X \to \mathbb{P}^1$ be an elliptic fibration of an elliptic surface X with $q(X) = 0$. Then

$$K_X \sim (p_g(X) - 1)F + \sum_i (m_i - 1)F_i^0$$

where F is any fibre of f , $F_i = m_i F_i^0$ all multiple fibres of multiplicity m_i .

Proof. For any non-singular fibre X_b we have

$$O_{X_b} \otimes \omega_X \simeq \omega_{X_b} \simeq O_{X_b} \quad .$$

Taking a sufficiently large number of distinct "general" points $b_1,\dots,b_r$ and considering the exact sequence

$$0 \longrightarrow \omega_X \longrightarrow \omega_X \otimes O(\sum_{i=1} X_{b_i}) \longrightarrow \bigoplus_{i=1}^{r} O_{X_{b_i}} \longrightarrow 0$$

we get

$$|K_X + \sum_{i=1}^{r} X_{b_i}| \neq 0 \quad .$$

If D is a divisor in the linear system above, we have

$$(D \cdot F) = 0 \ , \quad \text{for any fibre } F \ .$$

This implies that we can write

$$K_X \sim (\text{sum of fibres}) + \Gamma \quad ,$$

where $\Gamma \geq 0$ is contained in a union of fibres and does not contain fibres of f. Let Γ_0 be a connected component of Γ contained in the fibre X_b. If $X_b = \sum n_i E_i$ then

$$0 = (K_X X_b) = \sum n_i (K_X E_i)$$

and

$$0 = (X_b E_i) = \sum_{j \neq i} n_j (E_j E_i) + (E_i^2) \quad .$$

This shows that $(E_i^2) < 0$ if X_b is reducible, that implies that $(K_X E_i) = 2g(E_i) - 2 - (E_i^2) = 0$, since E_i cannot be an exceptional curve of the 1st kind. Hence, we have $(K_X E_i) = 0$.

Thus, if X_b is reducible, then

$$(\Gamma_0 \cdot E_i) = 0 \text{ , all components } E_i \text{ of } X_b .$$

Applying the main lemma we get that $\Gamma_0 = rX_b$, $r \in \mathbb{Q}$.

So, we have proved that

$$K_X \sim nF + \sum a_i F_i^0 , \quad 0 \leq a_i < m_i$$

and it remains to show that $n = p_g(X) - 1$, and $a_i = m_i - 1$.

For this we note, firstly, that the divisor $\sum a_i F_i^0$ is the fixed part of the linear system K_X. Indeed, any rational function belonging to the space $H^0(X, O_X(K_X))$ must be constant on the general fibre of f, and hence, it is induced by a rational function on $\mathbb{P}^1$. But then it is either regular on the divisor F_i^0, or has the pole of order multiple to m_i at F_i^0.

Thus we have

$$p_g = h^0(K_X) = h^0(nF) = n + 1$$

that proves the assertion about n.

Next, by Riemann-Roch

$$h^0(K_X + F_i^0) = 1 + p_g(X) > h^0(K_X)$$

and this shows that

$$K_X = F_i^0 \sim (n+1)F + \sum_i a_i' F_i^0 , \quad 0 \leq a_i' < m_i$$

(using again the arguments above) . This, obvioulsy, implies that $a_i + 1 = m_i$.

Corollary 1. For any minimal elliptic surface X

$$K_X^2 = 0 .$$

Furthermore, if $q(X) = 0$, then the plurigenus

$$P_n(X) = n(P_g(X) - 1) + \sum_{i=1}^{r} [n(m_i - 1)/m_i] + 1$$

Proof. The first assertion follows easily from the proof of the theorem. Indeed, we have proved without assumption $q(X) = 0$ that K_X is numerically equivalent to a rational linear combination of fibres.

To prove the second assertion, we use that

$$nK_X \sim n(p_g(X) - 1)F + \sum_{i=1}^{r} n(m_i - 1)F_{b_i}^0$$
$$(n(p_g(X) - 1) + \sum_{i=1}^{r} [n(m_i - 1)/m_i]F + \sum_{i=1}^{r} a_i F_{b_i}^0 ,$$

where $0 \leq a_i < m_i$. Again, using the arguments of the proof of the theorem, we get that $\sum a_i F_{b_i}^0$ equals the fixed part of $|nK_X|$. This, of course, proves the assertion.

Corollary 2. ([13]). An elliptic surface with $q = p_g = 0$ is rational if and only if its minimal elliptic fibration contains at most one multiple fibre.

In fact, $P_2(X) = 0$ implies that the number of multiple fibres $r \leq 1$. In another direction the assertion follows immediately.

Corollary 3. (Godeaux). Suppose that $q(X) = p_g(X) = 0$. Then

$$K_X \sim (r - 1)F - \sum_{i=1}^{r} F_{b_i}^0 ,$$

where F_{b_i}, $i = 1,\ldots, r$, are all multiple fibres.

Next, we want to compare the numerical invariants of an elliptic surface and its jacobian surface.

Proposition 6. Let $f : X \to B$ be an elliptic fibration. Denote by EP(Z) the topological Euler-Poincaré characteristic (in case $k \neq \mathbf{C}$, the field of complex numbers, we consider l-adic etale cohomology). Then

$$EP(X) = \sum_{b \in B} EP(F_b) .$$

For the proof we refer to [1], Ch. 4 ($k = \mathbf{C}$) or [12] (arbitrary k). Note that we use here the assumption $char(k) = 0$. In the general case there is some additional term depending on the wild ramification.

Corollary. Let X be a minimal elliptic surface, A its jacobian surface. Then

$$EP(X) = EP(A) , \quad p_a(X) = p_a(A) .$$

The first equality follows from propositions 1 and 2 , the second one follows from the first and the Noether formula.

Proposition 7. Let $f : X \to B$ be an elliptic fibration. Suppose that for some fiber X_b the reduced curve $X_{b,red}$ is singular. Then

$$q(X) = \text{genus}(B) .$$

Proof. The hypothesis implies that under the Albanese map alb: $X \to \text{Alb}(X)$ the fiber X_b goes to a point (since all of its components are rational curves). This shows that in the canonical commutative diagram

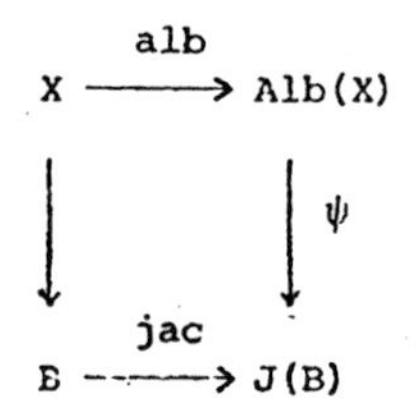

ψ is a finite surjective map and hence $\dim \text{Alb}(X) = \dim J(B) = \text{genus}(B)$.

Corollary. Suppose that the jacobian fibration $j : A \to B$ has a singular fiber. Then for any elliptic surface X with the jacobian surface equal to A we have

$$q(X) = q(A) .$$

Corollary. Let X be an elliptic surface with $q = p_g = 0$. Then its jacobian surface is rational. Conversely, any elliptic surface with rational jacobian surface has $q = p_g = 0$.

Thus all elliptic surfaces with $q = p_g = 0$ are obtained from rational jacobian elliptic surfaces by choice of some fibres of type 1^1h and some element of finite order of the Picard group of each of these fibrès.

All rational elliptic surfaces can be described with the help of so called <u>Halphen pencils</u> on the projective plane ([13]) . These are the pencils of curves of degree 3m with 9 multiple points of multiplicity m lying on a cubic. The case $m = 1$ corresponds to jacobian surfaces.

To find a place in the above classification of elliptic surfaces with $q = p_g = o$ for the Enriques surfaces constructed in Chapter 1 we note that for such surfaces $P_2 = 1$. In virtue of the first corollary to the theorem in §1 we get the following relation for the multiplicities m_i of multiple fibres

$$\sum_{i=1}^{r} (1-\frac{1}{m_i}) = 1 .$$

This, of course, can occur only in the case

$$r = 2, m_1 = m_2 = 2 .$$

Applying the formula for the canonical class of elliptic surfaces we see that, on the contrary, for any minimal elliptic surface X with $q = p_g = 0$ and two multiple fibres of multiplicity 2 we have $2K_X = 0$. Notice also that the following result holds:

Theorem (Enriques). Any algebraic surface X with $q = p_g = 0$ and $2K_X = 0$ is an elliptic surface.

The proof is too long to reproduce here (see [1], Ch. 9, and also [6]).

There is also a theorem (again due to Enriques) which states that any surface with $q = p_g = 0$, $2K = 0$ has a sextic surface as its birational model. Again the proof is too long to be reproduced here (see [1], Ch. 9 and also [2]). The particular form of this sextic passing through the edges of a tethraedron corresponds to a particular Enriques surface.

2. Torsion.

In this section we shall prove that any finite abelian group can be realized as the torsion group of an elliptic surface with $q = p_g = 0$.

Lemma. Suppose that D is a torsion divisor on an elliptic minimal surface X with $q = 0$. Then D is linearly equivalent to a rational linear combination of fibres of some elliptic fibration on X.

Proof. Since $h^2(K_X + D) = h^0(-D) = 0$, by Riemann-Roch we get

$$h^0(K_X + D) \geqq 1 + p_g(X) .$$

Let $D' \in |K_X + D|$. Since D' does not intersect a general fibre of any elliptic fibration (because K_X does not), it equals some linear combination of components of fibres. Moreover, D' does not intersect any component (because K_X does not). Applying the main lemma from §1 we get that D' is a rational linear combination of fibres. Thus, $D' - K_X \sim D$ is also a rational linear combination of fibres.

Theorem. Let $f : X \to B$ be a minimal elliptic fibration with $q(X) = 0$. Let $F_{b_i} = m_i F^0_{b_i}$, $i = 1, \ldots, r$ be all its multiple fibres. Then

$$\mathrm{Tors}(\mathrm{Pic}(X)) = \mathrm{Ker}\left(\bigoplus_{i=1}^{r} Z/m_i \xrightarrow{\psi} Z/m \right) ,$$

where $m = m_1 \ldots m_r$, $\psi(a_1, \ldots, a_r)) = \sum a_i \bar{m}_i \bmod m$, $\bar{m}_i = m/m_i$.

Proof. Using the lemma we may write any torsion divisor D in the form

$$D \sim \sum_{i=1}^{r} a_i F_{b_i}^{0} - lF ,$$

where $0 \leqq a_i < m_i$, F any non-multiple fibre.

Intersecting the both sides with some transversal curve C we obtain

$$0 = (D \cdot C) = \sum_{i=1}^{r} a_i (F \cdot C)/m_i - l(F \cdot C) ,$$

and hence

$$l = \sum_{i=1}^{r} a_i/m_i .$$

This shows that l is uniquely determined by a_i and, moreover,

$$(*) \qquad \sum_{i=1}^{r} a_i \bar{m}_i = lm \equiv 0 \mod m .$$

Now we know (see the proof of the theorem in §1) that the divisor $\sum a_i F_{b_i}^{0}$ is in the fixed part of any linear system containing it. Hence the coefficients a_i are determined uniquely by the divisor class of D . This shows that the map

$$\alpha : \mathrm{Tors}(\mathrm{Pic}(X)) \longrightarrow \mathrm{Ker}(\oplus Z/m_i \overset{\psi}{\to} Z/m)$$

$$D \longmapsto (a_1, \ldots, a_r)$$

is injective.

Now for any $(a_1,\dots,a_r)$ satisfying condition (*) the divisor

$$D = \sum_{i=1}^{r} a_i F_{b_i}^0 - 1F$$

has zero intersection with any transversal curve and any component of fibres. This shows that D is numerically equivalent to zero, and, hence, D is a torsion divisor. This proves the surjectivity of α .

Corollary 1. In notations above

$$\# \operatorname{Tors}(\operatorname{Pic}(X)) = \text{g.c.d.}(\bar{m}_1\dots, \bar{m}_r)$$

Corollary 2. For any finite abelian group G there exists an elliptic surface with $q = p_g = 0$ such that

$$\operatorname{Tors}(\operatorname{Pic}(X)) = G .$$

Proof. Applying Proposition 4 we may find such an elliptic surface with multiple fibres of any prescribed multiplicities.

Let

$$G = \bigoplus_{i=1}^{s}(Z/p_i^{n_1^{(i)}} \oplus \dots \oplus Z/p_i^{n_{k(i)}^{(i)}}) , \quad n_1^{(i)} \le \dots \le n_{k(i)}^{(i)}$$

be the primary decomposition of G . Consider a surface X with the following collection of multiplicities:

$$(p_1^{n_1^{(1)}},\dots,p_1^{n_{k(1)}^{(1)}}, p_1^{n_{k(1)}^{(1)}};\ \dots;\ p_s^{n_1^{(s)}},\dots,p_s^{n_{k(s)}^{(s)}},\ p_s^{n_{k(s)}^{(s)}}) .$$

Then applying the theorem we easily see that

$$\mathrm{Tor}(\mathrm{Pic}(X)) = G .$$

Corollary 3. ([14]) . There exists an elliptic surface with $q = p_g = 0$ which is not a rational surface and has no torsion divisors.

Just take a surface with multiple fibres of coprime multiplicities and apply Corollary 1 and Corollary 2 to the theorem of §1 .

§3. Fundamental group.

Here following to Kodaira [28] and Iithaka [25] we shall compute the fundamental group of an elliptic surface, over the field of complex numbers.

Let $f : X \to B$ be an elliptic fibration.

Lemma 1. Let $U \subset B$ be an open set such that the restriction $f_U : X_U \to U$ of f over U has no multiple fibres. Choose a point $p_0 \in X_U$ lying in a non-singular fibre. Then the following exact sequence holds

$$\pi_1(X_{f(p_0)}, p_0) \longrightarrow \pi_1(X_U, p_0) \longrightarrow \pi_1(U, f(p_0)) \longrightarrow 1$$

Proof. Consider the inclusion map $X_{f(p_0)} \hookrightarrow X_U$ and the projection map $f_U : X_U \to U$ and the correspondent homomorphisms of fundamental groups. Then the image of $\pi_1(X_{f(p_0)}, p_0)$ is clearly contained in the in the kernel of the second homomorphism, and we have to show that it coincides with the kernel and the second homomorphism is surjective. Restricted over sufficiently small U the map f is a differentiable 2-torus fibre bundle, and the corresponding sequence is the exact homotopy sequence. This obviously proves the surjectivity of the second homomorphism.

Let γ be a loop with the origin at p_0. Let X_{u_0} be a singular fibre of f_U, there exists a local section $D_{u_0} \to X_U$, D_{u_0} being a small disc centered at $f(p_0)$ (since X_{u_0} is not a multiple fibre):

Assuming that γ goes to zero under f_* it allows to deform γ to a loop on $X_{f(p_0)}$ keeping the point $p_0 \in \gamma$ fixed. This proves the lemma.

Next, let $D_1, \ldots, D_r$ be some open discs around the points $b_1, \ldots, b_r$ for which the fibre X_{b_i} is multiple of multiplicity m_i. Assume that over the punctured discs of D_i^* the morphism f is smooth. Let $U = B - D_1 - \ldots - D_r$, $X_U = f^{-1}(U)$, $V_i = f^{-1}(D_i)$, $V_i^* = f^{-1}(D_i^*)$.

We shall apply van Kampen's theorem to compute $\pi_1(X)$.

Let

δ, σ be some loops on $X_{f(p_0)}$ originated at p_0 which generates $\pi_1(X_{f(p_0)}; p_0)$

$t_1, \ldots, t_2$ be the loops on B starting at $f(p_0)$ and going around the points $b_1, \ldots, b_r$;

$a_1, \ldots, a_g$; $b_1, \ldots, b_g$ another loop originated at $f(p_0)$ which together with t_i generate $\pi_1(U; f(p_0))$;

Denote by t_i', a_i', b_i' some loops on X_U lying over t_i, a_i, b_i with the origin at p_0. Then assuming that a_i, b_i are chosen as the canonical generators of $\pi_1(U; f(p_0))$, we get the following.

<u>Lemma 2.</u> The group $\pi_1(X_U; p_0)$ is generated by δ, σ, $t_1', \ldots, t_r'$, $a_1', \ldots a_g', b_1', \ldots, b_g'$ (g = genus of B) with the following basic relations:

(i) $\delta\sigma = \sigma\delta$;

(ii) the group $\{ \delta, \sigma\}$ generated by δ, σ is normal in $\pi_1(X_U ; p_0)$;

(iii) $a_1'b_1'a_1'^{-1}b_1'^{-1}\ldots a_g'b_g'a_g'^{-1}b_g'^{-1}t_1'\ldots t_r' \in \{\delta, \sigma\}$;

(iv) some relation between δ and σ (may be trivial) .

This follows immediately from Lemma 1 and the known structure of $\pi_1(U ; f(p_0))$.

Choose some points p_i, $i = 1,\ldots, r$ lying over D_i^* and some loops δ_i^*, σ_i^* in the fibre $X_{f(p_i)}$ generating $\pi_1(X_{f(p_i)};p_i)$. Let $\bar{t}_i$ be a loop going around b_i and passing through $f(p_i)$, $\bar{t}_i'$ some loop on V_i^* lying over $\bar{t}_i$ which passes through p_i .

Lemma 3. The group $\pi_1(V_i^*, p_i)$ is generated by $\delta_i^*, \sigma_i^*, \bar{t}_i'$ with the following basic relations:

(i) $\delta_i^*\sigma_i^* = \sigma^*\delta^*$;

(ii) δ_i^* and σ_i^* generate a normal subgroup in $\pi_1(V_i^*; p_i)$;

(iii) $\bar{t}_i\delta_i^* = \delta_i^*\bar{t}_i$;

(iv) $\bar{t}_i\sigma_i^* = \delta_i^{*h_i}\sigma_i^*\bar{t}_i$, if X_{b_i} is of type $m_i^1h_i$.

Proof. Applying Lemma 1 we will prove the first assertion and find the first two relations. To obtain another pair of relations we will use the following description of V_i^* which is due to Kodaira [27] . There exists an unramified covering $F \to V_i^*$ whose covering transformation group is a cyclic group of order m_i . The space F is represented in the form

$$F = D_i^* \times \mathbb{C}/\Gamma \qquad , \text{ if } h_i = 0 ,$$

or

$$F = D_i^* \times \mathbb{C}^*/\Gamma \qquad , \text{ if } h_i > 0 ,$$

where in the first case Γ is the discontinuous group of analytic automorphisms

$$(z,\zeta) \longrightarrow (z,\zeta + n_1 j(z^{m_i}) + n_2) \ , \ n_1, n_2 \in \mathbb{Z}$$

($j(z^{m_i})$ is a holomorphic function of z^{m_i} with $\operatorname{Im} j(z^{m_i}) > 0$) . In the second case Γ is the infinite cyclic group of analytic automorphisms of $D_i^* \times \mathbb{C}^*$ generated by the automorphism

$$(z, w) \longmapsto (z, wz^{m_i h_i}) .$$

Identifying the universal covering space of D_i^* with the upper half plane $H = \{\tau \mid \operatorname{Im}(\tau) > 0 \}$ and the covering map with the exponential map $\tau \to \exp(2\pi i\tau)$, we get that in the both cases the universal covering space of V_i^* is equal to $H \times \mathbb{C}$ and the covering transformation group $\overline{\Gamma}$ may be described as follows: If $h_i = 0$, then $\overline{\Gamma}$ consists of analytic automoprhisms

$$(\tau, \zeta) \longmapsto (\tau + \frac{n_1}{m} , \zeta + n_3 j(\exp(2\pi i m_i \tau) + n_2) \ , \ n_1, n_j, n_3 \in \mathbb{Z} .$$

If $h_i > 0$, then $\overline{\Gamma}$ consists of analytic automorphisms

$$(\tau, \zeta) \longrightarrow (\tau + \frac{n_1}{m}, \zeta + n_2 + n_3 m_i h_i (\tau + \frac{n_1}{m})) .$$

Identifying in the usual way the loops originated at p_i with covering transformations, we may assume that

t_i corresponds to the element of $\overline{\Gamma}$ with $(n_1, n_2, n_3) = (1, 0, 0)$,

δ_i corresponds to the element of $\overline{\Gamma}$ with $(n_1, n_2, n_3) = (0, 1, 0)$,

σ_i corresponds to the element of $\overline{\Gamma}$ with $(n_1, n_2, n_3) = (0, 0, 1)$.

The relations (iii) and (iv) are verified now immediately.

To use van Kampen's theorem we consider homomorphisms

$$\pi_1(V_i^*, p_i) \to \pi_1(X_U, p_0) ,$$

which correspond to the natural inclusions $V_i^* \hookrightarrow X$ and to a choice of some paths connecting the points p_i and p_0 , and also the natural surjections

$$r_i : \pi_1(V_i^* ; p_i) \to \pi_1(V_i ; p_i) .$$

Applying the same arguments as in the proof of Proposition 2 from §1 we may assume that the cyclic covering F of V_i^* can be prolonged to an elliptic fibration over D_i , the cycle covering of D_i of degree m , with fibre of type ${}_1 1_{h_i}$ over the origin. This easily implies that

$$r_i(\overline{t}_i)^{m_i} \in \{p(\delta_i), p(\sigma_i)\} .$$

Moreover, if $h_i > 0$ we get that

$$p(\delta_i) = 0 .$$

Collecting everything together we obtain:

Theorem. The fundamental group $\pi_1(X)$ is generated by letters

$$\delta, \sigma, a_1, \ldots, a_g, b_1, \ldots, b_g, t_1, \ldots, t_r .$$

The basic relations are

i) $\delta \cdot \sigma = \sigma \cdot \delta$,

ii) $\{\delta, \sigma\}$ is a normal subgroup ,

iii) $t_i^{m_i} \in \{\delta, \sigma\}$; $i = 1, \ldots, r$,

iv) $a_1 b_1 a_1^{-1} b_1^{-1} \ldots a_g b_g a_g^{-1} b_g^{-1} t_1 \ldots t_r \in \{\delta, \sigma\}$.

v) some relation between δ and σ (may be trivial) .

Corollary. Let $f : X \to \mathbb{P}^1$ be an elliptic fibration. Then $\pi_1(X)$ is abelian if and only if it has at most 2 multiple fibres.

In fact, $\pi_1(X)$ has as its quotient the group $G(m_1, \ldots, m_r)$ given by generators $t_1, \ldots, t_r$ and relations

$$t_1^{m_1} = \ldots = t_r^{m_r} = t_1 \ldots t_r = 1 .$$

These groups are well known in the theory of automorphic functions. Namely, there exist natural representations of these groups as a discrete subgroup of the automorphism group of one of the three standard planes: the Riemannian $\mathbb{P}^1(\mathbb{C})$, the Euclidean $\mathbb{C}$, and the Lobachevsky $H = \{z \in \mathbb{C} \mid \mathrm{Im}(z) > 0\}$. We have (see [30]) that each of these cases corresponds to the sign of the number

$$e = \sum_{i=1}^{r} \frac{1}{m_i} - r + 2 \quad .$$

We have the case

$\mathbb{P}^1(\mathbb{C})$	iff $e > 0$ and iff $G(m_1,\ldots,m_r)$ is finite;
$\mathbb{C}$	iff $e = 0$ and iff $G(m_1,\ldots,m_r)$ is non-commutative nilpotent;
H	iff $e < 0$ and iff $G(m_1,\ldots,m_r)$ is infinite non-nilpotent .

Thus, $\pi_1(X)$ can be abelian only in the case $e > 0$. In this case, $G(m_1, m_2, m_3)$ is a finite subgroup of $SL(2,\mathbb{C})/\{\pm 1\}$, that is the rotation group of some regular polyhedron ($r = 3$) or a cyclic group ($r = 2$) . This, of course, proves the corollary.

Corollary. ([14]) . Let X be an elliptic surface with $q = p_g = 0$, which admits an elliptic fibration $f : X \to \mathbb{P}^1$ with exactly two multiple fibres of coprime multiplicity. Then X is a simply connected non-rational surface.

In fact, its fundamental group being abelian has to coincide with the homology group $H_1(X,\mathbb{Z})$. Since $q(X) = 0$, $H_1(X,\mathbb{Z}) = \mathrm{Tors}(H_1(X,\mathbb{Z})$. It remains to apply Corollary 1 of §2 .

Remark. In [14] the argument that $\pi_1(X)$ is abelian was not correct. So, in fact, it was proven there only that there exist non-rational surfaces with $q = p_g = 0$ with no torsion divisors. This was the original question of F. Severi.

CHAPTER III. SURFACES OF GENERAL TYPE

§1. Some useful lemmas.

Lemma 1. Let X be a scheme and T is a finite subgroup of the Picard group $\mathrm{Pic}(X)$. Then there exists a finite etale Galois covering $f : X_T \to X$ uniquely determined by the properties $T = \mathrm{Ker}(\mathrm{Pic}(X) \xrightarrow{f^*} \mathrm{Pic}(X_T))$ and the Galois group of f is isomorphic to the character group $\mathrm{Char}(T)$.

Proof. For any $\varepsilon \in T$ let $O_X(\varepsilon)$ be the corresponding invertible sheaf. The locally free sheaf $L = \bigoplus_{\varepsilon \in T} O_X(\varepsilon)$ has a natural structure of an O_X-Algebra corresponding to the isomorphisms $O_X(\varepsilon) \otimes O_X(\varepsilon') \to O_X(\varepsilon+\varepsilon')$. Put $X_T = \mathrm{Spec}(L)$. Then the projection $f : X_T \to X$ is finite and flat. It is also etale, since $\det(L) = \prod_{\varepsilon \in T} O_X(\varepsilon) = O_X$. The group $G = \mathrm{Char}(T)$ acts naturally on X_T multiplying each summand $O_X(\varepsilon)$ by $\chi(\varepsilon)$, $\chi \in G$. Clearly, the invariant subalgebra $L^G = O_X$, hence $X_T/G = X$ and f is a Galois covering. Assume that $\mathbb{L} \in \mathrm{Ker}(\mathrm{Pic}(X) \xrightarrow{f^*} \mathrm{Pic}(X_T))$. Then $f^*(\mathbb{L}) = O_{X_T}$ and $f_* f^*(\mathbb{L}) = f_*(O_{X_T}) = \bigoplus_{\varepsilon \in T} \mathbb{L} \otimes O_X(\varepsilon) = f_*(O_{X_T}) = \bigoplus_{\varepsilon \in T} O_X(\varepsilon)$. This implies that $\mathbb{L} \otimes O_X(\varepsilon) = O_X(\varepsilon')$ for some $\varepsilon' \in T$ and hence $\mathbb{L} = O_X(\varepsilon - \varepsilon') \in T$. The inclusion $T \subset \mathrm{Ker}$ is obvious.

To prove the uniqueness note that for any finite Galois covering $f : X' \to X$ with the Galois group G we have

$$\mathrm{Char}(G) \simeq \mathrm{Ker}(\mathrm{Pic}(X) \xrightarrow{f^*} \mathrm{Pic}(X')) .$$

This immediately follows from the Hochshild-Serre spectral sequence or from direct considerations.

Now $f_*(O_{X'})$ must split into eigen subsheaves corresponding to characters of G

$$f_*(O_{X'}) = \bigoplus_{\chi\in \mathrm{Char}(G)} f_*(O_{X'})_\chi$$

Let L_χ be the invertible sheaf corresponding to a character in virtue of the above identification of Char(G) with the subgroup of Pic(X) . Then L_χ being lifted onto X' is trivial, thus it is embedded into $f_*(O_{X'})$ and is isomorphic to one of its summands (namely, $f_*(O_{X'})_\chi$) . This shows that $X' = \mathrm{Spec}(f_*(O_{X'}))$ is isomorphic to X_T constructed above.

<u>Corollary</u>. In the above notations

$$H^i(X_T, O_{X_T}) = \bigoplus_{\varepsilon\in T} H^i(X, O_X(\varepsilon))$$

More generally, for any locally free sheaf L on X we have

$$H^i(X_T, f^*(L)) = \bigoplus_{\varepsilon\in T} H^i(X, L\otimes O_X(\varepsilon)) .$$

<u>Proof</u>. We have

$$f_*(O_{X_T}) = \bigoplus_{\varepsilon\in T} O_X(\varepsilon) ,$$

hence for any locally free L on X

$$f_*(f^*(L)) = f_*(O_{X_T})\otimes L = \bigoplus_{\varepsilon\in T} L\otimes O_X(\varepsilon) .$$

It remains to apply the Leray spectral sequence which degenerates because f is finite.

Lemma 2. (Bombieri [4]). Let F be a surface of general type with $q(F) = 0$, $m =$ Tors(F) the order of the torsion group. Then

$$p_g(F) \leq \frac{1}{2} K_F^2 + \frac{3}{m} - 1$$

and

$$p_g(F) \leq \frac{1}{2} K_F^2 - 1$$

if there exist a finite abelian unramified covering of F of irregularity at least one.

Proof. Let $f : \overline{F} \to F$ be the covering corresponding to the torsion group Tors(F) in virtue of Lemma 1. By the lemma of §2, Chapter 1 we know that

$$K_{\overline{F}}^2 = m\, K_F^2 ,$$

$$1 + p_a(\overline{F}) = m\,(1 + p_a(F)) .$$

Now apply the following classic Noether theorem (see [4], Th. 9):

$$p_g(\overline{F}) \leq \frac{1}{2} K_{\overline{F}}^2 + 2$$

and consider separately the two possible cases:

a) $q(\overline{F}) > 0$: Then $\mathrm{Pic}(\overline{F})$ contains a finite subgroup of any order n. Let $\overline{F}(n) \to \overline{F}$ be the corresponding etale covering.

We have

$$p_g(\overline{F}(n)) = n\ (1 + p_a(\overline{F})) + q(\overline{F}(n)) - 1 \leq \frac{1}{2} n\ K_F^2 + 2\ ,$$

dividing by n and letting $n \to \infty$ we get

$$1 + p_a(\overline{F}) \leq \frac{1}{2} K_F^2 \quad .$$

Now dividing by m we obtain

$$1 + p_a(F) = 1 + p_g(F) \leq \frac{1}{2} K_F^2 \quad .$$

b) $q(\overline{F}) = 0$: Then

$$m(1 + p_g(F)) = 1 + p_g(\overline{F}) \leq \frac{m}{2} K_F^2 + 3$$

and it suffices to divide both sides by m .

Lemma 3. Let F be a surface of general type. Then

$$q(F) \leq p_g(F) \quad .$$

Proof. By Noether's formula

$$1 - q(F) + p_g(F) = \frac{1}{12}(K_F^2 + c_2(F)) \ .$$

Since $K_F^2 > 0$ and $c_2(F) > 0$ (otherwise, F would be ruled, [4], Th. 13) we get the inequality.

Lemma 4. Let F be a surface of general type and D be a divisor numerically equivalent to mK_F, $m \geq 1$. Then

$$H^1(F, \mathcal{O}_F(D + K_F)) = 0 .$$

Proof. This immediately follows from the following Ramanujam's form of Kodaira's Vanishing theorem (C. Ramanujam, J. Indian Math. Soc., 38 (1974), 121-124) : Let X be a complete non-singular surface, L and invertible sheaf on X such that $(c_1(L)^2) > 0$ and $(c_1(L) \cdot C) \geq 0$ for any curve on X. Then $H^i(X, L^{-1}) = 0$ for $i = 0, 1$.

Corollary. The m-th plurigenus P_m of a surface of general type F is given by

$$P_m = \frac{1}{2} m(m - 1) K_F^2 + 1 + p_a(F) ,$$

in particular

$$P_2 = p^{(1)}(F) + p_a(F) .$$

Use Reimann-Roch and Lemma 4 applied to $D = (m - 1)K_F$.

Lemma 5. Let $f : X \to Y$ be a double covering of non-singular surfaces branched along a reduced curve $W \subset Y$. Then

(i) $f_*(\mathcal{O}_X) = \mathcal{O}_Y \oplus L$, $L^{\otimes 2} \simeq \mathcal{O}_Y(-W)$

(ii) $\omega_X = f^*(\omega_Y \otimes L^{-1})$.

Proof. The subsheaf O_Y is naturally identified with the subsheaf of $f_*(O_X)$ invariant under sheet-interchange. Since the characteristic is assumed to be zero (or at least prime to 2), this sheaf is a direct summand of $f_*(O_X)$, the complement being a sheaf L of anti-invariant sections. The sheaf $L^{\otimes 2}$ is obviously a subsheaf of the invariant subsheaf, that is O_Y , thus $L^{\otimes 2} \simeq J$ for some Ideal sheaf $J \subset O_Y$. This shows that X is isomorphic to the subscheme of the vector bundle $\mathbb{V}(L) = \operatorname{Spec}(\bigoplus_{n=0}^{\infty} L^{\otimes n})$ defined by the ideal $(L^{\otimes 2} - J)$. Now, the local arguments of the proof of Lemma 2, Ch. 1, §3 show that $J = O_Y(-W)$ and $\omega_X = f^*(\omega_Y \otimes L^{-1})$.

Corollary. Let F be an invertible sheaf on Y.. Then

$$H^0(X, f^*F) \simeq H^0(Y, F) \oplus H^0(Y, F \otimes L) .$$

In particular

$$H^0(X,\omega_X^{\otimes n}) \simeq H^0(Y, \omega_Y^{\otimes n} \otimes L^{-n}) \oplus H^0(Y,\omega_Y^{\otimes n} \otimes L^{-n+1}) .$$

§2. Numerical Godeaux surfaces.

By this we mean any surface of general type F with

$$p_g(F) = 0 \quad \text{and} \quad p^{(1)}(F) = 2 \ .$$

In virtue of Lemma 3 and corollary to Lemma 4 of §1 we get moreover that

$$q(F) = 0 \quad \text{and} \quad P_m(F) = \frac{1}{2}m(m-1) + 1 \ .$$

We will distinguish these surfaces by the value of its torsion group $\mathrm{Tors}(F)$. First of all, by Lemma 2 of 1 we have the following.

Proposition 1. If $m = \mathrm{Tors}(F)$ then

$$m \leq 6 \ .$$

For any abelian unramified covering $F' \to F$ we have

$$q(F') = 0 \ .$$

Proposition 2. (Bombieri). There are no numerical Godeaux surfaces with $\mathrm{Tors}(F) = 6$.

Proof. Assume that $\mathrm{Tors}(F) = \mathbb{Z}/2\mathbb{Z} \oplus \mathbb{Z}/3\mathbb{Z}$. Then there exists an unramified covering $F' \to F$ of order 2 with $\mathrm{Tors}(F') = \mathbb{Z}/3\mathbb{Z}$.

By the lemma of Chapter 1, §2 we have

$$p^{(1)}(F') = 3 \quad \text{and} \quad -q(F') + p_g(F') = 1 .$$

By proposition 1 $q(F') = 0$ and hence we obtain a surface with $p^{(1)} = 3$, $p_g = 1$, $q = 0$ and the torsion group $\mathbb{Z}/3\mathbb{Z}$. However this contradicts Theorem 15 of [4] .

Remark. Since the previous proof is a simple application of Theorem 15 of [4], which in its turn is proved using other non-trivial results of [4], it is better to give an independent proof. As suggested by Miles Reid we can argue as follows.

Let Y be the covering of X corresponding to the group of torsion of order 6. Then $p_g(Y) = 5$, $K_Y^2 = 6 = 2p_g(Y) - 4$. Now we will use

Lemma (E. Horikawa). Let Y be a surface of general type with $(K_Y^2) = 2p_g(Y) - 4$. There $|K_Y|$ is an irreducible linear system whose general member is a hyperelliptic curve.

Proof. Suppose that

$$|K_Y| = |C| + F$$

where F is a fixed part. Assuem that $|C|$ is composed of a pencil, say $C \sim a[C_0]$, where $a > 1$ and $[C_0]$ is an irreducible pencil. Then $p_g(Y) \leq a + 1$ and the equality holds if $[C_0]$ is linear (i.e. $\dim H^0(Y, O(C_0)) = 2$) . We have $K_Y \cdot F \geq 0$, therefore $K_Y^2 \geq aK_Y \cdot F$ and since $C_0^2 \geq 0$ we get $K_Y \cdot C_0 \geq 2$, because $K_Y^2 \geq 2$.

Hence

$$K_Y^2 \geq 2a \geq 2p_g - 2$$

and we have a contradiction. Thus we may assume that $|C|$ is not composed of a pencil.

Now the analysis of the proof of Noether's inequality $p_g(Y) \leq \frac{1}{2} K_Y^2 + 2$ (see [4], p. 209) shows that in the case of the equality $|K_Y|$ is an irreducible non-singular curve C of genus $g = (K_Y^2) + 1$.

Now the exact sequence

$$0 \longrightarrow O_Y \longrightarrow O_Y(K_Y) \longrightarrow O_C(K_Y \cdot C) \longrightarrow 0$$

shows that $\dim H^0(C, O_C(K_Y \cdot C)) = p_g(Y) - 1$. Let D denotes the restriction of $|K_Y|$ on C . Then $2D \sim K_C$ and $2 \dim H^0(C, O_C(D)) = $ $= 2p_g(Y) - 2 = K_Y^2 + 2 = \deg D + 2$. Now by a classical Clifford's theorem on special divisors it follows that C is hyperelliptic (see, for example, H. Martens. J. Réine Angen. Math. 233, (1968), 89–100).

After we have proven the lemma the argument is very simple. If σ is an automorphism of the covering $Y \to X$ then σ acts freely on Y and hence on a general member C of $|K_Y|$. But this is obviously impossible (any automorphism of a hyperelliptic curve has a fixed point)..

Lemma. (Reid [39]). Let F be a minimal numerical Godeaux surface. Then

(i) For any non-zero $g \in \mathrm{Tors}(F)$ there exists a unique positive divisor $D_g \in |K_F + g|$;

(ii) if $g \neq g'$ then D_g and $D_{g'}$ have no common components;

(iii) if g, g' and g'' are distinct non-zero elements of $\mathrm{Tors}(F)$ then D_g, $D_{g'}$ and $D_{g''}$ do not meet.

Proof. (i) By Riemann-Roch

$$h^0(K_F + g) = 1 + h^1(K_F + g) - h^2(K_F + g) .$$

By Serre's duality, $h^2(K_F + g) = h^0(-g) = 0$, since $g \neq 0$. By the same reason, $h^1(K_F + g) = h^1(-g) = 0$ in virtue of the corollary to Lemma 1, §1 and Proposition 1 .

(ii) If one of D_g or $D_{g'}$ is irreducible the result is obvious. Suppose that

$$D = D_g = C + \sum n_i C_i \ , \quad D' = D_{g'} = C' + \sum n_i' C_i'$$

is the decomposition into irreducible components with C and C' chosen so that $(D \cdot C) = (K_F \cdot C') = 1$ (recall that $(D \cdot K_F) = (D' \cdot K_F) = K_F^2 = 1)$.

If $C = C'$ then $D = D'$, because there are no relations between fundamental curves (that is, curves with no intersection with K_F) other than equality ([4], Prop. 1) .

Let E be the common part of D and D', then $E^2 < 0$ and even, since it is a positive combination of fundamental curves. Thus $(D - E)^2 = D^2 - 2(D \cdot E) + E^2 = 1 + E^2 \leq -1$. But

$$(D - E)^2 = (D - E)(D' - E)$$

must be non-negative, since $D - E$ and $D' - E$ have no common components.

(iii) Since $K_F^2 = 1$ each two D_g and $D_{g'}$, $g \neq g'$ meet transversally at a non-singular point for both curves. The fact that three distinct D_g, $D_{g'}$ and $D_{g''}$ meet at a point is equivalent to the fact that $O_F(D_g - D_{g'})$ being restricted on $D_{g''}$ is isomorphic to the structure sheaf of $D_{g''}$. Write the exact sequence

$$0 \to O_F(D_g - D_{g'} - D_{g''}) \to O_F(D_g - D_{g'}) \to O_{D_{g''}} \to 0$$

and the corresponding cohomology sequence

$$H^0(F, O_F(D_g - D_{g'})) \to H^0(D_{g''}, O_{D_{g''}}) \to H^1(F, O_F(D_g - D_{g'} - D_{g''})) \quad .$$

Since $D_g - D_{g'}$ is a non-zero torsion divisor, the first term is zero. By duality, the third term is equal to $h^1(\varepsilon)$ for some torsion divisor. That is also zero (see the proof of (i)). This contradicts the non-triviality of the middle term.

Proposition 3. **(Bombieri-Catanese, Reid). There are no numerical Godeaux surfaces with $\mathrm{Tors}(F) = \mathbb{Z}/2\mathbb{Z} \oplus \mathbb{Z}/2\mathbb{Z}$.**

Proof. **Let F be such a surface. Then we have the three distinct**

non-zero torsion divisors of order 2. Let D, D' and D'' be the three divisors constructed in Reid's lemma. Then the divisors $2D$, $2D'$ and $2D''$ belong to the linear system $|2K_F|$ and by the property (iii) they cannot be members of a pencil. Thus, $\dim |2K_F| \geq 2$. However, we know that $P_2(F) = \dim |2K_F| + 1 = 2$. This contradiction proves the assertion.

Remark. The proof of Bombieri-Catanese [5] uses other more elaborate arguments. The proof from [32] is not complete. Thus, we have the following possible cases:

$$\mathrm{Tors}(F) = \{0\} \ ; \ \mathbb{Z}/2\mathbb{Z} \ ; \ \mathbb{Z}/3\mathbb{Z} \ , \ \mathbb{Z}/4\mathbb{Z} \ \text{ and } \ \mathbb{Z}/5\mathbb{Z} \ .$$

We know examples of surfaces with $\mathbb{Z}/5\mathbb{Z}$ (the Godeaux surfaces of §2, Chapter I). Let us show that these are essentially all examples of such surfaces. The proof below is due to Miles Reid [39].

Let $\bar{F}$ be the unramified covering of order 5 corresponding to the torsion group $\mathrm{Tors}(F)$. Then by the corollary to Lemma 1 of §1 we have

$$H^0(\bar{F}, O_{\bar{F}}(mK_{\bar{F}})) = \bigoplus_{g \in \mathrm{Tors}} H^0(F, O_F(mK_F + g)) \ .$$

We know form Reid's lemma (i) that $h^0(K + g) = 1$, $g \neq 0$. Let x_1, x_2, x_3, x_4 be non-zero elements corresponding to the four non-zero elements of $\mathrm{Tors}(F)$. We may consider them as elements of $H^0(\bar{F}, O_{\bar{F}}(K_{\bar{F}}))$ generating this space. Since by Reid's lemma the x_i's have no common zero on F, therefore on $\bar{F}$ they define a morphism $f : F \to \mathbb{P}^3$.

Since $K_{\overline{F}}^2 = 5$ and the degree of f must divide 5 we get that f is birational onto a surface F' of degree 5. This quintic F' must be a normal surface, since the arithmetic genus of its hyperplane sections coincides with the genus of its inverse images (=canonical divisors) on $\overline{F}$. Thus F' coincides with the canonical model of $\overline{F}$ and as such has only double rational points as singularities.

The group $G = \mathrm{Char}(\mathrm{Tors}(F)) = \mathbb{Z}/5\mathbb{Z}$ acting on $\overline{F}$ acts by functoriality on the canonical model $F' = \mathrm{Proj}(\bigoplus_{m=0}^{\infty} H^0(\overline{F}, O_{\overline{F}}(mK_{\overline{F}})))$ multiplying x_i by some ζ^i (ζ a 5-th root of unity). Thus F is "almost" the quotient of a quintic by $\mathbb{Z}/5\mathbb{Z}$. More exactly, the canonical model of F is isomorphic to such quotient.

We refer to [11] and [32] for the study of pluricanonical maps of numerical Godeaux surfaces. Also in [32] it can be found the facts concerning the moduli space of surfaces with $\mathrm{Tors} = \mathbb{Z}/5\mathbb{Z}$.

Surfaces with $\mathrm{Tors}(F) = \mathbb{Z}/4\mathbb{Z}$ (Reid-Miyaoke).

To construct such surfaces we will pull ourselves by shoe-strings. Assume that such surface F exists. As for the Godeaux surfaces we consider the elements $x_i \in H^0(F, O_F(K_F + g_i))$, where g_1, $g_2 = g_1^2$, $g_3 = g_1^3$ are non-zero elements of $\mathrm{Tors}(F)$. Then x_1x_3 and x_2^2 form a basis for $H^0(F, O_F(2K_F + g_2))$ (their linear independence follows from Reid's lemma). Let y_1 and y_3 be sections of $H^0(F, O_F(2K_F + g_1))$ and $H^0(F, O_F(2K_F + g_3))$ respectively such that (x_2x_3, y_1) and (x_1x_2, y_3) form bases.

Proposition. (Reid). The above elements x_i, y_i generate the pluricanonical ring $A(\overline{F}) = \bigoplus_{m=0}^{\infty} H^0(\overline{F}, O_{\overline{F}}(mK_{\overline{F}})) = \bigoplus_{\substack{m=0 \\ g \in \mathrm{Tors}}}^{\infty} H^0(F, O_F(mK_F + g))$

of the surface $\overline{F}$ which is the unramified covering of F corresponding to the torsion group $\mathrm{Tors}(F)$. There are two basic relations of degree 8 between these generators.

Proof. The monomials

$$x_1^4, x_2^4, x_3^4, x_1^2x_3^2, x_1x_2^2x_3, y_1y_3, y_1x_1x_2, y_3x_3x_2 \in H^0(F, O_F(4K_F)) .$$

However, by the corollary to Lemma 4, §1 we find that

$$h^0(4K_F) = 7 .$$

Thus there is a linear dependence between these 8 monomials, which we will write

$$f_0(x_1, x_2, x_3, y_1, y_3) = 0 .$$

In the same way the 8 monomials,

$$x_1^2x_2^2, x_3^2x_2^2, x_1^3x_3, x_1x_3^3, y_1^2, y_3^2, x_1x_2y_3, x_3x_2y_1 \in H^0(F, O_F(4K_F + g_2))$$

and $h^0(4K_F + g_2) = 7$. Hence we have the second relation

$$f_1(x_1, x_2, x_3, y_1, y_3) = 0 .$$

Both these relations of degree 4 considering x_i, y_i as elements of the graded canonical ring $A(\overline{F}) = \bigoplus_{m=0}^{\infty} H^0(\overline{F}, O_{\overline{F}}(mK_{\overline{F}}))$.

Next, let

$$B = \mathbb{C}[X_1,X_2,X_3,Y_1,Y_3]/(f_0,f_1)$$

be the quotient polynomial ring. Grade B by the condition $\deg(X_i) = 1$, $\deg(Y_i) = 2$, then we have the morphism of graded algebras

$$\psi : B \to A(\bar{F}) \ , \quad X_i \mapsto x_i \ , \ Y_i \mapsto y_i \quad .$$

The proposition is equivalent to the assertion that ψ is an isomorphism.

Now, the Poincaré function (compare [15])

$$P_B(t) = \sum \dim B_i \ t^i = \frac{(1-t^4)^2}{(1-t)^3(1-t^2)^2} = \frac{(1+t^2)^2}{(1-t)^3} =$$

$$= \sum \left(\frac{(1+2)(i+1)}{2} + i(i-1) + \frac{(i-2)(i-3)}{2}\right) t^i =$$

$$= \sum (2i(i-1)+4)t^i \quad .$$

In virtue of the formula for $P_i(\bar{F})$ this coincides with

$$P_{A(\bar{F})}(t) = \sum P_i(\bar{F}) \ t^i \quad .$$

Thus, it suffices to check that ψ is injective.

If ψ is not injective then the image of the rational map

$$\psi : M = \mathrm{Proj}(A(\overline{F})) \longrightarrow V = \mathrm{Proj}(B)$$

will be a proper closed subscheme of V.

Let j be the embedding $V \hookrightarrow \mathbb{P}^7$ corresponding to the surjection $\mathbb{C}[B_2] \to B^{(2)} = \bigoplus_{i=0}^{\infty} B_{2i}$, $a : \overline{F} \to M$ the canonical map of $\overline{F}$ onto its canonical model M. The composition

$$\overline{F} \longrightarrow M \longrightarrow V \longrightarrow \mathbb{P}^7$$

is easily to be seen coincides with the 2-canonical map

$$\varphi_{2K_{\overline{F}}} : \overline{F} \to \mathbb{P}^7 .$$

In virtue of Reid's lemma $\varphi_{K_{\overline{F}}}$ is regular (see the analogous argument in the previous case of the Godeaux surfaces), thus $\varphi_{2K_{\overline{F}}}$ is also regular. This shows that $\tilde{\psi}$ is in fact a morphism.
Let $\overline{V} = \varphi_{2K_{\overline{F}}}(\overline{F})$. By our assumption, $\overline{V}$ is a proper closed subscheme of V. Since $\overline{V}$ spans $\mathbb{P}^7$ its degree is at least 6. Since $(2K_{\overline{F}})^2 = 16$ and $|K_{\overline{F}}|$ has no fixed part it implies that $\overline{V}$ is a surface and $\deg \overline{V} = 8$ or 16. Moreover, in the first case, $\varphi_{2K_{\overline{F}}}$ defines a 2-sheet covering

$$g : \overline{F} \longrightarrow \overline{V},$$

and in the second case g is a birational morphism. Since $\deg j(V) = 16$

(this follows from the equality of the Poincare functions for $A(\overline{F})$ and B) we get that in the second case $\overline{V} = V$. So, we may assume that $\varphi_{2K_{\overline{F}}}$ is a 2-sheeted covering onto its image $\overline{V}$. Let $C \in |K_{\overline{F}}|$ be a non-singular curve, the map $g|_C$ equals the canonical map of C and since it is 2-sheeted C must be a hyperelliptic curve and $g|_C$ its hyperelliptic involution. Now, notice that the canonical map $\varphi_{K_{\overline{F}}}$ also factors through $\tilde{\psi}$ and hence through g. Then $K_{\overline{F}}$ cuts out on C a g_4^1 which is composed with hyperelliptic g_2^1. This implies that $K_{\overline{F}|C}$ is not a complete linear system. But the latter contradicts the vanishing of $H^1(\overline{F}, O_{\overline{F}})$.

Corollary. Let F be a numerical Godeaux surface with $\mathrm{Tors}(F) = \mathbb{Z}/4\mathbb{Z}$, $\overline{F}$ its unramified covering corresponding to the torsion group. Then the canonical model $\overline{M}$ of $\overline{F}$ is isomorphic to a weighted complete intersection $V_{4,4}(1,1,1,2,2)$. The action of the group $\mathrm{Char}(\mathbb{Z}/4\mathbb{Z}) = \mu_4$ on $\overline{M}$ is induced by the action of this group on the weighted projective space $\mathbb{P}(1,1,1,2,2)$ which multiplies the first three coordinates by ζ, ζ^2, ζ^3 accordingly and the fourth and the fifth coordinate by ζ, ζ^3 accordingly (ζ a primitive 4-th root of 1). The canonical model M of F is obtained by dividing $\overline{M}$ by this action.

This corollary prompts to us the way to construct F. For this one may take a non-singular $F = V_{4,4}(1,1,1,2,2)$ invariant under the above action on $\mathbb{P}(1,1,1,2,2)$ and not containing the fixed point of this action. Using the general properties of weighted complete intersection (which are quite analogous to the ones of usual non-singular complete intersections) we find (see, for example, [15]):

$$q(\overline{F}) = 0 \ , \quad \omega_{\overline{F}} = \mathcal{O}_{\overline{F}}(4+4-1-1-1-2-2) = \mathcal{O}_{\overline{F}}(1) \ ,$$

$$p_g(\overline{F}) = \dim H^0(\overline{F}, \mathcal{O}_{\overline{F}}(1)) = 3, \ P_2(\overline{F}) = \dim H^0(F, \mathcal{O}_{\overline{F}}(2)) = 8 \ ,$$

$$p^{(1)}(\overline{F}) = P_2(\overline{F}) - P_a(\overline{F}) = 5 \ .$$

Dividing F by the free action of μ_4 we get the surface F with

$$q(F) = p_a(F) = p_g(F) = 0 \ , \quad p^{(1)} = 2 \ .$$

Notice also that we have $\pi_1(\overline{F}) = 0$ and thus

$$\pi_1(F) = \mathrm{Tors}(F) = \mathbb{Z}/4\mathbb{Z} \ .$$

An explicit example of $V_{4,4}(1,1,1,2,2)$ with the properties above:

$$x_0^4 + x_1^4 + x_2^4 + x_3 x_4 = 0$$

$$x_0^2 x_1^2 + x_2^2 x_1^2 + x_3^2 + x_4^2 = 0 \ .$$

For a more general example see [32] .

<u>Surfaces with $\mathrm{Tors}(F) = \mathbb{Z}/3\mathbb{Z}$.</u>

Here the same method of Miles Reid shows that the covering $\overline{F}$ of such surface F is embedable into the weighted projective space $\mathbb{P}(1,1,2,2,2,3,3)$, unfortunately, not as a complete intersection. There are not any explicit constructions of $\overline{F}$ (the example in [39] does not work) and, thus, the question of the existence of such surfaces <u>F is still open* .</u>

* see Epilogue.

Surfaces with $\mathrm{Tors}(F) = Z/2Z$ (Campedelli-Kulikov-Oort).

The main idea here belongs to Campedelli, who proposed to construct a surface with $p^{(1)} = 2$ as a double plane branched along a 10th order curve with 5 triple points of type $x^3 + y^6 = 0$ and an ordinary 4-ple point. Unfortunately, his construction of such a curve is false (see below). Victor Kulikov (non-published) proposed to modify the Campedelli curve, taking the union of two conics and two cubics such that one of the cubics has a double point, both conics pass through this point and touch both the cubics at other points. Oort gave an explicit construction of this configuration ([35]): Let $W = C_1 \cup C_2 \cup D_1 \cup D_2$, where

$$C_1 : y^2 + (x-t)(2x-2y-3t) = 0$$

$$C_2 : y^2 + (x-t)(2x+2y-3t) = 0$$

$$D_1 : y^2t + x(x-t)(x-3t) = 0$$

$$D_2 : [(y^2t + x(x-t)(x-3t))(2t-x)+(x^2-3xt+3t^2)^2]/t = 0$$

It is easily checked that

$$C_1 \cap D_1 = 2P_1 + 2P_2 + 2P_5 ,$$

$$C_2 \cap D_1 = 2P_3 + 2P_4 + 2P_5 ,$$

$$C_1 \cap D_2 = 2P_1 + 2P_2 + 2P_6 ,$$

$$C_2 \cap D_2 = 2P_3 + 2P_4 + 2P_6 ,$$

$$C_1 \cap C_2 = 3P_5 + P_6 ,$$

$$D_1 \cap D_2 = 2P_1 + 2P_2 + 2P_3 + 2P_4 + P_7 .$$

where

$$P_1 = (x,y,t) = (\frac{3+\sqrt{-3}}{2}, \frac{3+\sqrt{-3}}{2}, 1); \; P_2 = (\frac{3-\sqrt{-3}}{2}, \frac{3-\sqrt{-3}}{2}, 1) ;$$

$$P_3 = (\frac{3+\sqrt{-3}}{2}, \frac{3+\sqrt{-3}}{2}, 1), \quad P_4 = (\frac{3-\sqrt{-3}}{2}, -\frac{3-\sqrt{-3}}{2}, 1),$$

$$P_5 = (1,0,1), \; P_6 = (\frac{3}{2},0,1), \; P_7 = (0,1,0) ,$$

the point P_6 is an ordinary double point of D_2, and the combination of the points above is considered as a divisor on any non-singular curve taking part in the intersection.

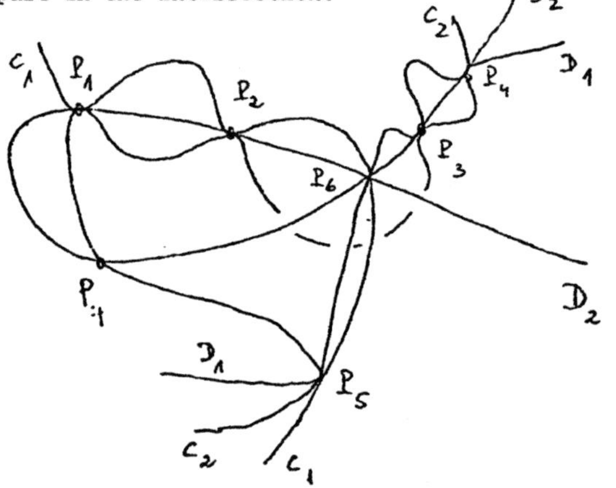

Let F be the minimal non-singular model of the double plane branched along the curve W.

<u>Assertion 1:</u>

$$p_g(F) = 0, \quad p^{(1)}(F) = 2 .$$

Proof. This is similar to the proof used at the construction of the classical Campedelli surface from Chapter 1, §3 .

Let $p : X \to \mathbb{P}^2$ be the minimal resolution of singular points of the branch curve W . Then the strict inverse transform of W

$$p^{-1}(W) \sim 10p^*(L) - 3\sum_{i=1}^{5} S_i - 6\sum_{i=1}^{5} S_i' - 8S_5'' - 4S_6 - 2S_7 ,$$

where L is a line on $\mathbb{P}^2$,

$$p^{-1}(P_i) = S_i + S_i' , \; i = 1,\ldots, 4; \; p^{-1}(P_5) = S_5 + S_5' + S_5'' ;$$

$$p^{-1}(P_i) = S_i , \; i = 6, 7 ;$$

with $S_i^2 = -2$, $1 \leq i \leq 5$; $S_i^2 = -1$, $i = 6, 7$; $S_i'^2 = -2$, $S_5''^2 = -1$.

Let $r : F' \to X$ be the double covering of X branched along the divisor $p^{-1}(W) + \sum_{i=1}^{5} S_i$, then

$$\begin{aligned} K_{F'} &\sim r^*(K_X) + \tfrac{1}{2}r^*(p^{-1}(W) + \sum_{i=1}^{5} S_i) \\ &\sim r^*(p^*(-3L) + \sum_{i=1}^{7} S_i + 2\sum_{i=1}^{5} S_i' + 3S_5'') + \\ &+ \tfrac{1}{2}r^*(10p^*(L) - 2\sum_{i=1}^{5} S_i - 6\sum_{i=1}^{5} S_i' - 8S_5'' - 4S_6 - 2S_7) \\ &\sim r^*(p^*(2L) - \sum_{i=1}^{5} S_i' - S_5'' - S_6) . \end{aligned}$$

Assume that $D \in |K_{F'}|$, then we see from above and corollary to Lemma 5, §1 that $D = r^*(D')$, where $D' \in |2p^*(L) - \sum_{i=1}^{5} S_i' - S_5'' - S_6|$ and hence equals the proper inverse image under p of a conic passing through the points $P_1, \ldots, P_6$. However, obviously these points are not situated on a conic. This shows that $|K_{F'}| = \emptyset$ and thus

$$p_g(F) = 0 .$$

Since r is branched along S_i, $i=1,\ldots,5$ and $p^{-1}(C_i)$, $i=1,2$, we see that

$$r^*(S_i) = 2\bar{S}_i) \ , \ r^*(p^{-1}(C_i)) = 2\bar{C}_i'$$

for some curves S_i and $\bar{C}_i'$ on F' . Also, we have

$$\bar{S}_i^2 = \frac{1}{4}(r^*(S_i)^2) = \frac{1}{4}(2S_i^2) = \frac{1}{4}(-4) = -1$$

$$\bar{C}_i'^2 = \frac{1}{4}(r^*(p^{-1}(C_i))^2) = \frac{1}{4}(2(p^{-1}(C_i))^2) = \frac{1}{4}(-8) = -2 .$$

This shows that $\bar{S}_i$ are exceptional curves of the 1st kind. Let $\sigma : F' \to F$ be the blowing down of all $\bar{S}_i$. We will show that F is the minimal model of F' . We have

$$\begin{aligned} 2K_F &\sim \sigma_*(r^*(4p^*(L) - 2\sum_{i=1}^{5} S_i' - 2S_5'' - 2S_6)) \\ &\sim \sigma_*(r^*(p^{-1})(C_1)) + r^*(p^{-1}(C_2)) + 2r^*(S_5') + 4r^*(S_5'')) \\ &\sim \sigma_*(2\bar{C}_1' + 2\bar{C}_2' + 2r^*(S_5') + 4r^*(S_5'')) \end{aligned}$$

and hence

$$2K_F \sim 2\bar{C}_1 + 2\bar{C}_2 + 2\bar{S}_5' + 4\bar{S}_5'' \ ,$$

where we put

$$\sigma_*(\bar{C}_1' = \bar{C}_i, \ \ \sigma_*(r^*(S_5')) = \bar{S}_5', \ \sigma_*(r^*(S_5'')) = \bar{S}_5'' \ .$$

Assuming that E is an exceptional curve of the 1st kind on F, we get that $(E\cdot 2K_F) = -2$ and hence E coincides with one of the curves $\bar{C}_i$, $\bar{S}_5'$ or $\bar{S}_5''$. However, we saw above that $\bar{C}_i^2 = \bar{C}_i'^2 = -2$ and also $\bar{S}_5'^2 = r^*(S_5')^2 + 1 = 2\,S_5'^2 + 1 = -4+1 = -3$, $\bar{S}_5''^2 = r^*(S_5'')^2 = -2.$

Now

$$p^{(1)}(F) = K_F^2 + 1 = \frac{1}{4}(2K_F)^2 + 1 = \frac{1}{4}(-8-8-12-32+32+16+16\,) + 1 = 2$$

and the assertion is proven.

Assertion 2.

$$\mathrm{Tors}(F) = \mathbb{Z}/2\mathbb{Z}\quad .$$

Proof. In the proof of Assertion 1 we have found already a torsion divisor of order 2, this is

$$K_F - \bar{C}_1 - \bar{C}_2 - \bar{S}_5' - 2\bar{S}_5''\quad .$$

In virtue of the analysis of the torsion of numerical Godeaux surfaces we know that $\mathrm{Tors}(F) = \mathbb{Z}/2\mathbb{Z}$ or $\mathbb{Z}/4\mathbb{Z}$. Let us exclude the second possibility.

Assume that g is a torsion divisor of order 4. Consider the involution δ of F corresponding to its rational projection onto $\mathbb{P}^2$. If $\delta^*(g) \sim g$, then $2g \sim 0$, since there are no torsion divisors on $\mathbb{P}^2$. Thus, $\delta^*(g) \sim -g$, because δ defines an automorphism of the torsion group $\mathbb{Z}/4\mathbb{Z}$. Let D_g be the unique curve from $|K_F + g|$. Then

$$D_g + \delta^*(D_g) = D_g + D_{-g} \in |2K_F|\quad .$$

The bicanonical system $|2K_F|$ is a pencil generated by the two curves

$$2\overline{C}_1 + 2\overline{C}_2 + 2\overline{S}_5' + 4\overline{S}_5''$$

and

$$\overline{D}_2 + \overline{H} + \overline{S}_5'' + 2\sigma_*(2^*(S_7)) \sim \sigma_*(2^*(2p^*(L) - 2\sum_{i=1}^{4} S_i' - 2S_6 - S_7)) +$$

$$+ \sigma_*(2^*(p^*(L) - 2S_5' - 3S_5'' - S_7)) +$$

$$+ \overline{S}_5'' + 2\sigma_*(2^*(S_7)) \sim \sigma_*(2^*(4p^*(L) -$$

$$- 2\sum_{i=1}^{5} S_i' - 2S_5'' - 2S_6)) \sim 2K_F \quad .$$

We see that $|2K_F|$ has the fixed component, namely $\overline{S}_5''$, which has to be contained in both D_g and D_{-g}. However, by Reid's lemma the curves D_g and D_{-g} has no common components. This contradiction proves the assertion.

Remark. Campedelli proposed to construct the branch curve W as the union of 3 conics C_1, C_2, C_3 and a quartic D such that C_1 and C_2 are bitangent to C_3, touch each other at a point, D has a node at one of the two ordinary intersection points of C_1 and C_2, passes through the five contact points of the conics with the same tangent direction (see [9]) .

The arguments similar to the one used above show that the bicanonical system of the corresponding double plane is equal to the inverse image of the pencil of quartics on $\mathbb{P}^2$ touching D at the points of contact with $C_1 \cup C_2 \cup C_3$ and having a node at the node of D. Considering the two curves from this pencil $C_1 + C_2$ and D we will find two torsion divisors of order 2. This contradicts Proposition 3. Thus the Campedelli construction does not exist.

Surfaces with Tors(F) = 0 .

There are no examples of such surfaces. Maybe it is worth to consider a version of the example above with the branch curve W equal to the union of two conics and two cubics forming the following configuration (Kulikov):

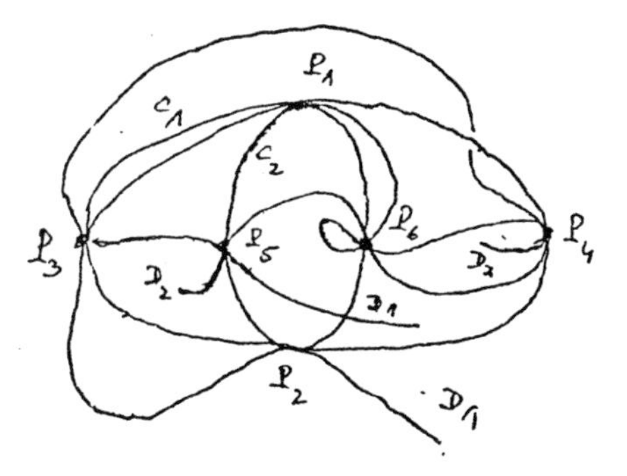

where C_1 and C_2 are conics, and D_1, D_2-cubics.

Arguing as above we would show that the bicanonical system is equal to the inverse image of the pencil of quartics passing through $P_1,\ldots,P_5$ with the same tangent direction as W and having a node at P_6 . It is seen that there are no members of this pencil composed of components of W . This easily proves that there are no torsion elements of order 2 .

Of course, the existence of this configuration is not easy to justify.

3. Numerical Campedelli surfaces.

These are surfaces with $p_g = 0$ and $p^{(1)} = 3$. They are distinguished by the order m of its torsion gorup. It was proved by Beauville [3] and Reid that $m \leq 10$. Here we exhibit examples of numerical Campedelli surfaces with $m = 2, 4, 7$ and 8. There are no examples of such surfaces with other possible value of m*, moreover there are no examples of numerical Campedelli surfaces with $Tors(F) = \mathbb{Z}/4\mathbb{Z}$.

a) Classical Campedelli surfaces. For them we already know (Chapter 1, §3) that $Tors(F) \supset (\mathbb{Z}/2\mathbb{Z})^3$. We will prove now that we have the equality.

Proposition (Miyaoke [32], Reid [39]). Let $r : \bar{F} \to F$ be the unramified covering of the classical Campedelli surface corresponding to the subgroup $T = (\mathbb{Z}/2\mathbb{Z})^3$ of the torsion group $Tors(F)$. Then the canonical system $K_{\bar{F}}$ defines the birational morphism of $\bar{F}$ onto the intersection of 4 quadrics in $\mathbb{P}^6$.

Proof. We know (Chpater III, §1) that

$$H^0(\bar{F}, O_{\bar{F}}(nK_{\bar{F}})) = \bigoplus_{g \in T} H^0(F, O_F(nK_F + g)).$$

Let us show that

$$h^0(K_F+g) = \dim H^0(F, O_F(K_F+g)) = 1, \text{ for all non-zero } g \in T.$$

Since $h^0(2K_F) = 3$, we get that $h^0(K_F+g) \leq 2$. If we have the equality, then $|2K_F|$ is composed of the pencil $\quad +g|$.

---

* see Epilogue.

Considering the restriction of $|K_F + g|$ onto $\bar{S}_1$, we see that this pencil has a base point on $\bar{S}_1$. This shows that $|2K_F|$ has also this point as its base point. However, the curves $2\bar{C}_3 + 2\bar{C}_2 + 2\bar{S}_3 + 2\bar{S}_4$ and $2\bar{C}_1 + 2\bar{C}_3 + 2\bar{S}_5 + 2\bar{S}_6$ from $|2K_F|$ intersect $\bar{S}_1$ at two distinct points. This contradiction proves the needed assertion.

Denote the lements of T by 000, 100, 110, 010, 001, 011, 101, and 111 . Let

$$x_0 \in H^0(K_F + 100),\ x_1 \in H^0(K_F + 010),\ x_2 \in H^0(K_F + 001)\ ,$$
$$x_3 \in H^0(K_F + 011)\quad x_4 \in H^0(K_F + 101),\ x_5 \in H^0(K_F + 110)\ ,$$
$$x_6 \in H^0(K_F + 111)$$

be non-zero sections.

Clearly, $r^*(x_i) = y_i$, $i=0,\ldots,6$, generate $H^0(\bar{F},O_{\bar{F}}(K_{\bar{F}}))$. All squares x_i^2 belong to $H^0(F,O_F(2K_F))$ and, since $h^0(2K_F) = 3$, there must be 4 relations among them. This shows that there are 4 relations between y_i^2 in $H^0(\bar{F},O_{\bar{F}}(K_{\bar{F}}))^2$. Now we can find explicitly these relations. We know that the bicanonical system $|2K_F|$ is represented by the net of quartics

$$\lambda_1 C_1 C_2 + \lambda_2 C_1 C_3 + \lambda_3 C_2 C_3$$

(in notation of Ch. I, §3). Up to a permutation we easily find that

x_1^2 corresponds to C_1C_2

x_2^2 " " C_2C_3

x_3^2 " " C_1C_3

$$\begin{array}{lll} x_4^2 & \text{corresponds to} & C_1\ell_2^2 \\ x_5^2 & \text{"} \quad \text{"} & C_2\ell_2^2 \\ x_6^2 & \text{"} \quad \text{"} & C_3\ell_3^2 \\ x_7^2 & \text{"} \quad \text{"} & D \end{array}$$

where ℓ_1 (resp. ℓ_2, resp. ℓ_3) is the line through the points P_5 and P_6 (resp. P_3 and P_4, resp. P_1 and P_2) .

This gives the following relations among y_i

$$y_1^2 = ay_3^2 + by_4^2$$
$$y_2^2 = cy_3^2 + dy_6^2$$
$$y_3^2 = cy_1^2 + ey_5^2$$
$$y_7^2 = fy_1^2 + gy_2^2 + hy_3^2$$

for some non-zero constants $a, b, \dots, g, h$.

Thus we obtain that the canonical image $\Phi_{\overline{K}}(\overline{F})$ is contained in the complete intersection V of the four quadrics given above. It is easily checked that V has only isolated singular points (in fact 24 double ordinary points) and hence being a complete intersection is an irreducible surface. This implies that $\Phi_{\overline{K}}(\overline{F}) = V$ if only $\dim \Phi_{\overline{K}}(\overline{F}) = 2$. Assume that $\Phi_{\overline{K}}(\overline{F})$ is a curve. Then its normalization X is isomorphic to the projective line $\mathbb{P}^1$ (since $g(\overline{F}) = 0$ in view of the corollary to Lemma 1, Ch. III, §1 and the remark above concluding that $h^0(K_{\overline{F}} + \varepsilon) = 1$ for any $\varepsilon \in T$) . Clearly the group $T = (\mathbb{Z}/2)^3$ acts faithfully on $\mathbb{P}^7 = \mathbb{P}(H^0(\overline{F}, \mathcal{O}_{\overline{F}}(K_{\overline{F}}))$ and hence on the image $\Phi_{\overline{K}}(\overline{F})$. this shows that T is isomorphic to a subgroup of $\mathrm{Aut}(\mathbb{P}^1)$, but this is impossible.

Thus we obtain that

$$V = \mathrm{Proj}(\bigoplus_{m=0}^{\infty} H^0(\overline{F}, O_{\overline{F}}(K_{\overline{F}}))^m) = \cdot(\overline{F})$$

is a complete intersection of four quadrics.

<u>Remark</u>. Computing the Poincare function of the canonical ring $A(\overline{F}) = \bigoplus_{m=0}^{\infty} H^0(\overline{F}, O_{\overline{F}}(mK_{\overline{F}}))$ we see that it coincides with the Poincare function of its subring $\bigoplus_{m=0}^{\infty} H^0(\overline{F}, O_{\overline{F}}(K_{\overline{F}}))^m$. This shows that these rings are isomorphic and V is the canonical model of $\overline{F}$. In particular V has exactly 24 double ordinary points corresponding to the inverse images of the three (-2)-curves on F : $\overline{C}_1$, $\overline{C}_2$ and $\overline{C}_3$. Also we get that the canonical model of F is the quotient of V by the group $(\mathbb{Z}/2)^3$. In this way it is easily to get the moduli space of the classical Campedelli surfaces. It is a unirational variety of dimension 6 (look at the coefficients of the four equations of V above). See the details in [32].

<u>Corollary</u>. Let F be a classical Campedelli surface. Then

$$\mathrm{Tors}(F) = \pi_1(F) = (\mathbb{Z}/2\mathbb{Z})^3 .$$

In fact, the surface F obtained as the unramified covering of F corresponding to the subgroup $(\mathbb{Z}/2\mathbb{Z})^3 \subset \mathrm{Tors}(F)$ is simply-connected (because it is isomoprhic to a minimal resolution of double rational points of a complete intersection).

b) <u>Godeaux' surfaces</u>. These surfaces were constructed by Godeaux as the quotients of suitable intersections of four quadrics in $\mathbb{P}^6$ by cyclic group of order 8 acting freely ([20]).

Consider four quadrics given by the equations:

$$a_1x_0x_6 + a_2x_1x_5 + a_3x_2x_4 + a_4x_3^2 = 0$$

$$b_1x_0^2 + b_2x_4^2 + b_3x_2x_6 + b_4x_3x_5 = 0$$

$$c_1x_1^2 + c_2x_5^2 + c_3x_0x_2 + c_4x_4x_6 = 0$$

$$d_1x_2^2 + d_2x_6^2 + d_3x_0x_4 + d_4x_1x_3 = 0$$

where a generator of $G = \mathbb{Z}/8\mathbb{Z}$ acts on the intersection X of these quadrics by the formulas:

$$(x_0,x_1,x_2,x_3,x_4,x_5x_6) \longrightarrow (x_0,\zeta x_1,\zeta^2x_2,\zeta^3x_3,\zeta^4x_4,\zeta^5x_5,\zeta^6x_6)$$

where $\zeta = \exp(2\ i/8)$.

The same argument as in the case of classical Godeaux surfaces shows that the quotient X/G is a numerical Campedelli surface with

$$\mathrm{Tors}(\mathrm{Pic}(X/G) = \pi_1(X/G) = \mathbb{Z}/8\mathbb{Z} .$$

c) Godeaux-Reid surfaces. These are also quotients of the intersection of four quadrics by other groups of order 8 ([39]). First, consider the group $G = (\mathbb{Z}/2\mathbb{Z})^3$. Define the action of G on $\mathbb{P}^6$ by the formulas:

$$g_1 : (x_0,x_1,x_2,x_3,x_4,x_5,x_6) \longrightarrow (-x_0,x_1,x_2,x_3,-x_4,-x_5,-x_6)$$

$$g_2 : (x_0,x_1,x_2,x_3,x_4,x_5,x_6) \longrightarrow (x_0,-x_1,x_2,-x_3,x_4,-x_5,-x_6)$$

$$g_3 : (x_0,x_1,x_2,x_3,x_4,x_5,x_6) \longrightarrow (x_0,x_1,-x_2,-x_3,-x_4,x_5,-x_6)$$

It is clear that for any fixed point (i.e. a point with non-trivial isotropy subgroup) at least three of its coordinates must be zero. This shows that G acts freely on the surface given by the equations

$$\sum a_i x_i^2 = \sum b_i x_i^2 = \sum c_i x_i^2 = \sum d_i x_i^2 = 0 \quad ,$$

where all minors of maximal order of the matrix

$$\begin{pmatrix} a_0 & \cdots & a_6 \\ b_0 & \cdots & b_6 \\ c_0 & \cdots & c_6 \\ d_0 & \cdots & d_6 \end{pmatrix}$$

are non-zero.

Second, consider the group $G = \mathbb{Z}/2\mathbb{Z} \oplus \mathbb{Z}/4\mathbb{Z}$. Let $g_1 = (1,0)$, $g_2 = (0,1)$ be its generators. Define the action of G on $\mathbb{P}^6$ by the formulas ($\zeta = e^{\pi i/2}$) :

$$g_1 : (x_0,x_1,x_2,x_3,x_4,x_5,x_6) \longrightarrow (-x_0,x_1,x_2,x_3,-x_4,-x_5,-x_6)$$

$$g_2 : (x_0,x_1,x_2,x_3,x_4,x_5,x_6) \longrightarrow (x_0, x_1,-x_2, \zeta^3 x_3,\zeta x_4 .-x_5,\zeta^3 x_6)$$

Now notice that any fixed point is fixed either under g_1 or under g_2^2 . Thus, the set of the fixed point in $\mathbb{P}^6$ with respect to the action of G is the set

$$F = \{x_1=x_2=x_3=0\} \cup \{x_0=x_4=x_5=x_6=0\} \cup \{x_1=x_3=x_4=x_6=0\} \cup \{x_0=x_2=x_5=0\} \ .$$

This shows that the surface X given by the equations

$$a_0x_0^2 + a_1x_2^2 + a_2x_5^2 + a_3x_1x_3 + a_4x_4x_6 = 0$$

$$b_0x_0^2 + b_1x_2^2 + b_2x_5^2 + b_3x_1x_3 + b_4x_4x_6 = 0$$

$$c_0x_1^2 + c_1x_3^2 + c_2x_6^2 + c_3x_0x_5 + c_4x_4^2 = 0$$

$$d_0x_1^2 + d_1x_3^2 + d_2x_6^2 + d_3x_0x_5 + d_4x_4^2 = 0$$

is easily can be chosen not passing through F . Since it is obviously G-invariant we may consider the quotient X/G , which is a numerical Campedelli surface with

$$\pi_1(X/G) = \mathbb{Z}/2\mathbb{Z} \oplus \mathbb{Z}/4\mathbb{Z} \quad .$$

The last example is more interesting [40] . Let $Q = \{\pm 1, \pm i, \pm j, \pm k\}$ be the quaternion group. Consider tis action on $\mathbb{P}^6$ by the formulas:

$$-1 : (x_0,x_1,x_2,x_3,x_4,x_5,x_6) \longrightarrow (x_0,x_1,x_2,-x_3,-x_4,-x_5,-x_6)$$

$$i : (x_0,x_1,x_2,x_3,x_4,x_5,x_6) \longrightarrow (-x_0,x_1,x_2,x_4,-x_3,x_6,-x_5)$$

$$j : (x_0,x_1,x_2,x_3,x_4,x_5,x_6) \longrightarrow (x_0,-x_1,x_2,x_5,-x_6,-x_3,x_4)$$

$$j : (s_0,x_1,x_2,x_3,x_4,x_5,x_6) \longrightarrow (-x_0,-x_1 + x_2,x_6,x_5,-x_4,-x_3)$$

Since $g^2 = -1$ for all $g \neq 1$, any fixed point is fixed by -1 . This shows that the set of fixed points

$$F = \{x_0=x_1=x_2=0\} \cup \{x_3=x_4=x_5=x_6=0\} \quad .$$

Now, the surface X given by the equations:

$$a_0x_0x_1 + a_1x_3x_4 + a_2x_5x_6 = 0$$

$$b_0x_1x_2 + b_1x_3x_5 + b_2x_4x_6 = 0$$

$$c_0x_0x_2 + c_1x_3x_6 + c_2x_4x_5 = 0$$

$$d_0x_0^2 + d_1x_1^2 + d_2x_2^2 + d_3(x_3^2+x_4^2+x_5^2+x_6^2) = 0$$

is G-invariant and obviously can be chosen to be non-singular and not passing through F . Taking the quotient $V = X/G$ we obtain a numerical Campedelli surface with

$$\pi_1(V) = Q_8 \ , \quad \mathrm{Tors}(V) = \mathbb{Z}/2\mathbb{Z} \oplus \mathbb{Z}/2\mathbb{Z} \ .$$

d) *Surfaces with Tors = Z/7Z* . It is proven by Godeaux [21] and Reid [39] that if such surface F exists then the canonical model $\overline{F}$ of its covering corresponding to the torsion group is given by seven cubical equations in $\mathbb{P}^5$. More precisely, it is shown by Reid that the surface $X \subset P^5$ given by the equations

$$x_2^2x_0+x_4^2x_3+x_5^2x_1+x_0x_1x_3-x_0^2x_4-x_1^2x_2-x_3^2x_5-x_2x_4x_5 = 0$$

$$-x_4^3+x_5^2x_2+x_0^2x_5+x_0x_2x_3-x_1^2x_3-x_2^2x_1 = 0$$

$$-x_2^3+x_4^2x_5+x_2^2x_5-x_1x_5x_0-x_3^2x_0-x_5^2x_3 = 0$$

$$-x_0^3+x_3^2x_1+x_2^2x_3-x_2x_1x_4+x_5^2x_4-x_1^2x_5 = 0$$

$$-x_5^3+x_2^2x_4-x_3^2x_2+x_3x_4x_1-x_0^2x_1+x_4^2x_0 = 0$$

$$-x_3^3+x_1^2x_0-x_4^2x_1+x_0x_4x_5+x_2^2x_5+x_0^2x_2 = 0$$

$$-x_1^3+x_0^2x_3-x_5^2x_0-x_5x_3x_2-x_4^2x_2+x_3^2x_4 = 0$$

is a very good candidate to be such surface $\overline{F}$. It is certainly invariant with respect to the involution δ of $\mathbb{P}^5$

$$(x_0,x_1,x_2,x_3,x_4,x_5) \longrightarrow (x_0,\zeta^2x_1,\zeta^3x_2,\zeta^4x_3,\zeta^5x_4,\zeta^6x_5) \ ,$$

where $\zeta = \exp(2\pi i/7)$. Also, this involution acts freely on X .

It has the same Hilbert polynomial as $\bar{F}$. The only thing that has to be proven is that X is non-singular and canonically embeded.

e) Campedelli-Oort-Kulikov surfaces. The history here is the same as in the case of similar surfaces with $p^{(1)} = 2$. Kulikov proposed to modify the classical Campedelli surface replacing the branch curve W by another curve also of the 10th order. More precisely, the new W is constructed as the union $W = E \cup F \cup C \cup D$, where E and F are non-singular cubics, C and D are conics, which intersect each other according to the following picture:

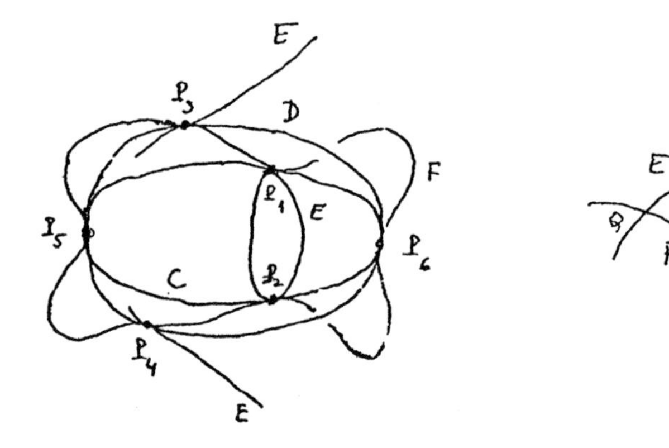

Oort gave the explicit equations (in affine coordinates):

$$E : y^2 + x(x^2 + x + 2) = 0$$

$$F : (x + 1)^2(x - 3)^2 - (x + 3)(y^2 + x(x^2 + x + 2)) = 0$$

$$C : y^2 - x^2 + x = 0$$

$$D : y^2 + 7x^2 - 7x = 0$$

The same arguments as in the case of all other double planes considered above show that the bicanonical system of the surface equals the inverse image of the linear system of quartics passing through P_i with the same tangent direction as W. Also, in the same manner it can be shown that the minimal non-singular model of the corresponding double plane is a numerical Campedelli surface. The curves $C \cup D$, $C \cup 2L$, $D \cup 2L'$, where L (resp. L') is the line given by the equation $x + 1 = 0$ (resp. $x - 3 = 0$) determine the bicanonical divisors effectively divisible by 2. Thus, they define three torsion divisors of order 2, whose sum is, in fact, linearly equivalent to zero. This shows that

$$\mathrm{Tors}(F) \supset (\mathbb{Z}/2\mathbb{Z})^2 .$$

It is easy to see that there are no more torsion divisors of order 2. Applying Beauville's estimate of #Tors we get that

$$\mathrm{Tors}(F) = (\mathbb{Z}/2\mathbb{Z})^2 \quad \text{or} \quad (\mathbb{Z}/2) \oplus \mathbb{Z}/4\mathbb{Z} .$$

Unfortunately, I cannot see how to exclude the second possibility. But it is conjectured that it can be done.

Remark. We have two different constructions of surfaces with $\mathrm{Tors} = (\mathbb{Z}/2\mathbb{Z})^3$, these are the classicla Campedelli surfaces and the Godeaux-Reid surfaces. It is easy to see (using the proposition from this section) that the Godeaux-Reid surface is a deformation of the classical Campedelli surface (see the details in [36]).

4. Burniat's surfaces

These surfaces were constructed in [7,8] as certain (2,2)-covers of the projective plane. The linear genus $p^{(1)}$ takes value 3, 4, 5, 6, and 7 for them. Later this construction was reproduced in a modern way by C. Peters [37] . Here I give some other version of this construction which allows to compute the torsion group.*

First, we consider a minimal rational elliptic surface $V \to \mathbb{P}^1$ with two exceptional fibres $F_0 = 2E_0 + E_1 + E_2 + E_3 + E_4$ and $F_0' = 2E_0' + E_1' + E_2' + E_3' + E_4'$ of type 1_0^* (see Ch. II, §1). We also suppose that there exist 4 sections S_1, S_2, S_3, S_4 nonintersecting each other with the properties:

$$(S_i E_i) = (S_i E_i') = 1 ,$$

$$2S_i + E_i + E_i' \sim 2S_j + E_j + E_j' .$$

To construct such a surface V one may consider the ruled surface $\mathbb{F}_2$, that is a $\mathbb{P}^1$-bundle over $\mathbb{P}^1$ with a section s_0 for which $(s_0^2) = -2$, an elliptic pencil on it generated by the curves $2s_0 + \ell_1 + \ell_2 + \ell_3 + \ell_4$ and $2s$, s being any section nonintersecting s_0 and ℓ_i any four distinct fibres of $\mathbb{F}_2$. The minimal resolution of the base points of this pencil $s \cap \ell_i$ provides the needed elliptic surface V.

Next, let F_1 and F_2 be any two distinct non-singular fibres of V, consider the pencil P generated by the divisors $F_1+2S_3+E_3+E_3'$ and $F_2+2S_4+E_4+E_4'$. It is easily seen that P has 2 base points

---

* See Epilogue

of multiplicity 2, namely, $Q_1 = F_1 \cap S_4$ $Q_2 = F_2 \cap S_3$. Moreover, F_1 (resp. F_2) touches non-singular curves of the pencil at Q_1 (resp. Q_2).

Let D_1 and D_2 be two curves of $\mathcal{P}$ without common components. Consider the following five possible cases (it will be shown later that all of them can be realized):

A) D_i are both non-singular;

B) $D_1 = E_1 + D_1'$, where D_1' is non-singular, D_2 as in A);

C) D_1 as in B), $D_2 = E_1' + D_2'$, where D_2' is non-singular;

D) $D_1 = E_1 + E_2' + D_1'$, where D_1' is non-singular, D_2 as in C);

E) D_1 as in D), $D_2 = E_2 + E_1' + D_2'$, where D_2' is non-singular.

The following properties are easily checked:

$$(D_i^2) = 4 \ , \quad (D_i K_V) = -(D_i F) \doteq -2 \quad (F \text{ any fibre}) \ ,$$

D_1' touches D_2' at Q_1 and Q_2, $(D_1' \cdot D_2') = 4$,

D_i does not meet any of E_j or E_j',

$(D_i' \cdot E) = 2$, where E denotes any other irreducible component of D_i.

The Burniat surfaces will be constructed as minimal non-singular models of the double covering of V branched along the curve W, where in each of the cases A)-E) the curve W is as follows:

A). $W = D_1+D_2+F_1+F_2+\sum_{i=1}^{4} E_i+\sum_{i=1}^{4} E_i' \sim 6F+4S_1+2E_1+2E_1'-2E_0-2E_0'$,

B) $W = D_1'+D_2+F_1+F_2+\sum_{i=2}^{4} E_i+\sum_{i=1}^{4} E_i' \sim 6F+4S_1+2E_1'-2E_0-2E_0'$,

C) $W = D_1'+D_2'+F_1+F_2+\sum_{i=2}^{4} E_i+\sum_{i=2}^{4} E_i' \sim 6F+4S_1-2E_0-2E_0'-2E_2'$,

D) $W = D_1'+D_2'+F_1+F_2+\sum_{i=3}^{4} E_i+\sum_{i=2}^{4} E_i' \sim 6F+4S_1-2E_0-2E_0'-2E_2-2E_2'$,

E) $W= D_1'+D_2'+F_1+F_2+\sum_{i=3}^{4} E_i+\sum_{i=3}^{4} E_i' \sim 6F+4S_1-2E_0-2E_0'-2E_2-2E_2'$:

The following pictures represent W in cases A) and D) :

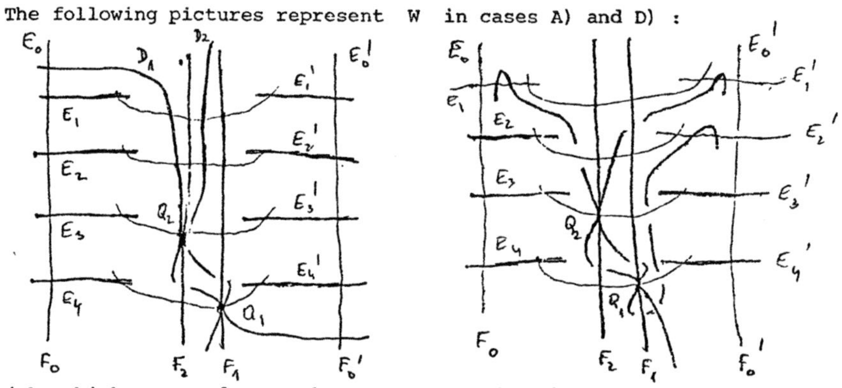

(the thick curves denote the components of W) .

To get a minimal non-singular model of this double covering we proceed as in the case of the classical Campedelli surfaces. Let $\sigma : V' \to V$ be the birational morphism which blows up the curves R_i and R_i' at the points Q_i $(i=1,2)$, where we assume that

$$(R_i^2) = -1 , \quad (R_i'^2) = -2 .$$

Then the divisor

$$p^{-1}(W) + R'_1 + R'_2 \sim p^*(W) - 2R_1' - 2R_2' - 6R_1 - 6R_2$$

is 2-divisible and non-singular. Thus we may form a double covering $r : X' \to B'$ branched along this divisor which will be a non-singular model of X .

To compute $K_{X'}$ we use the formula of Ch. I, §2:

$$K_{X'} = r^*(K_{V'}) + \frac{1}{2} r^*(p^{-1}(W) + R_1' + R_2') =$$

$$= r^*(p^*(-F) + R_1' + R_2' + 2R_1 + 2R_2) +$$

$$+ \frac{1}{2} r^*(p^*(W) - 2R_1' - 2R_2' - 6R_1 - 6R_2) \sim$$

$$\sim r^*(-p^*(F) + R_1' + R_2' + 2R_1 + 2R_2 + 3p^*(F) +$$

$$+ 2p^*(S_1) - p^*(E_0') + B - R_1' - R_2' - 3R_1 - 3R_2) \sim$$

$$\sim r^*(p^*(2F + 2S_1 - E_0 - E_0') + B - R_1 - R_2) ,$$

where

$$\begin{aligned} B &= r^*(p^*(E_1 + E_1') , && \text{in case A)} ,\\ &= r^*(p^*(E_1')) , && \text{in case B)} ,\\ &= 0 , && \text{in case C)} ,\\ &= -r^*(p^*(E_2')) , && \text{in case D)} ,\\ &= -r^*(p^*(E_2 + E_2')), && \text{in case E)} . \end{aligned}$$

Now notice that $p^*(R_i')$ are exceptional curves of the 1st kind taken with multiplicity 2 . The same is true also for $r^*(p^*(E_i))$ or $r^*(p^*(E_1'))$ if r is branched along $p^*(E_i)$ or $p^*(E_i')$. Let $\sigma : X' \to X$ be the blowing down these exceptional curves. Put $\overline{D} = \sigma_*(r^*(p^*(D)))$ for any divisor D on V , and also $\overline{R}_i = \sigma_*(r^*(R_i))$. Then, we get

$$K_X \sim 2\overline{F} + 2\overline{S}_1 - \overline{E}_0 - \overline{E}_0' - \overline{R}_1 - \overline{R}_2 + B ,$$

where

$$\hat{B} = \begin{cases} 0 & , \text{ in cases A), B), C)} , \\ -\bar{E}_2' & , \text{ in case D)} , \\ -\bar{E}_2 - \bar{E}_2' & , \text{ in case E)} . \end{cases}$$

Since

$$\begin{aligned} \bar{F} \sim 2\bar{E}_0 \sim 2\bar{E}_0' \quad & , \text{ in case A)} , \\ 2\bar{E}_0 + \bar{E}_1 \sim 2\bar{E}_0' \quad & , \text{ in case B)} , \\ 2\bar{E} \;\; + \bar{E}_1 \sim 2\bar{E}_0' + \bar{E}_1' \quad & , \text{ in case C)} , \\ 2\bar{E}_0 + \bar{E}_1 \sim 2\bar{E}_0' + \bar{E}_1' + \bar{E}_2' \quad & , \text{ in case D)} , \\ 2\bar{E}_0 + \bar{E}_1 + \bar{E}_2 \sim 2\bar{E}_0' + \bar{E}_1' + \bar{E}_2' \quad & , \text{ in case E)} , \end{aligned}$$

and

$$\bar{F}' \sim 2\hat{F}_0 + 2\bar{R}_1 \sim 2\hat{F}_0' + 2\bar{R}_2 \quad ,$$

where $\bar{F}_0 = 2\hat{F}_0$, $\bar{F}_0' = 2\hat{F}_0'$, we get

$$\begin{aligned} 2K_X & \sim 4\bar{F} + 4\bar{S}_1 - 2\bar{E}_0 - 2\bar{E}_0' - 2\bar{R}_1 - 2\bar{R}_2 + 2\hat{B} \\ & \sim 2\hat{F}_0 + 2\hat{F}_0' + 4\bar{S}_1 , \text{ in case A)} , \\ & \sim 2\hat{F}_0 + 2\hat{F}_0' + 4\bar{S}_1 + \bar{E}_1 , \text{ in case B)} , \\ & \sim 2\hat{F}_0 + 2\hat{F}_0' + 4\bar{S}_1 + \bar{E}_1 + \bar{E}_1' , \text{ in case C)} , \\ & \sim 2\hat{F}_0 + 2\hat{F}_0' + 4\bar{S}_1 + \bar{E}_1 + \bar{E}_1' - \bar{E}_2' \sim 2\hat{F}_0 + 2\hat{F}_0' + 2\bar{S}_1 + 2\bar{S}_2 + \bar{E}_2 , \text{ in case D)} , \\ & \sim 2\hat{F}_0 + 2\hat{F}_0' + 4\bar{S}_1 + \bar{E}_1 + \bar{E}_1' - \bar{E}_2 - \bar{E}_2' \sim 2\hat{F}_0 + 2\hat{F}_0' + 2\bar{S}_1 + 2\bar{S}_2 , \text{ in case E)} \end{aligned}$$

This implies

a) $K_X^2 = \frac{1}{4}((2K_X)^2) = 6$, in case A) ,
$= 5$, in case B) ,
$= 4$, in case C) ,
$= 3$, in case D) ,
$= 2$, in case E) .

b) X is non-rational (since $2K_X$ is positive) .

c) X is a minimal model (since for any exceptional curve of the 1st kind C $(2K_X C) < 0$ and this implies that C is one of the curves $\hat{F}_0$, $\hat{F}'_0$, $\bar{E}_i$ or $\bar{E}'_i$, but it is easily checked that neither of them is an exceptional curve of the 1st kind).

It remains to show that

$$p_g(X) = 0 .$$

For simplicity we will prove it only in the case A) . In other cases the proof is similar.

Suppose that $|K_X| \neq \emptyset$. Then taking its inverse transform on X' we get

$$r^*(p^*(2F-E_0-E'_0+\sum E_i+\sum E'_i+2S_1) \geq r^*(R_1) + r^*(R_2) .$$

This implies that

$$p^*(2F-E_0-E'_0+\sum E_i+\sum E'_i+2S_1) \geq R_1 + R_2 .$$

This means that there exists a positive divisor

$$D \in |2F - E_0 - E_0' + \sum E_i + \sum E_i' + 2S_1| = |E_0 + E_0' + \sum E_1 + \sum E_i' + 2S_1|$$

which passes through the points Q_1 and Q_2.

Now notice that

$$|D| \supset |E_1 + E_1' + 2S_1| + |E_0 + E_0' + \sum_{i=2}^{4} E_i + \sum_{i=2}^{4} E_i'|$$

moreover, $D^2 = 0$, and $(D \cdot K_V) = -2$. If $\dim |D| > 1$ then for the moving part $|D'|$ of $|D|$ we must have $(D'^2) > 0$. Thus $|D|$ has some fixed part which clearly consists of components of $E_0 + E'_0 + \sum_{i=2}^{4} E_i + \sum_{i=2}^{4} E_i'$ (since $|E_1 + E_1' + 2S_1|$ is an urreducible pencil of rational curves). However, it can be seen that adding any of these components to $E_1 + E_1' + 2S_1$ does not increase the self-intersection index. This shows that $|E_1 + E_1' + 2S_1|$ is, in fact, equal to the moving part of $|D|$. Thus, since the fixed part of D does not contain the points Q_1 and Q_2, we have to show that there are no curves in $|E_1 + E_1' + 2S_1|$ passing through Q_1 and Q_2. But this is easy, because the only curve linearly equivalent to $E_1 + E_1' + 2S_1$ passing thrugh Q_1 is the curve $E_3 + E_3' + 2S_3$ which does not pass through Q_2.

The only thing hanging on us is the proof of the existence of the cases A)-E). Of course, for A) it is easy, since the general member of the pencil P is non-singular. To construct other cases we use a representation of V as a double plane which comes from the inversion involution of the general elliptic fibre of V. Dividing V by this involution we get the surface Z obtained from the quardic

$\mathbb{P}^1 \times \mathbb{P}^1$ by blowing up 8 points, the four of them P_1, P_2, P_3, P_4 are situated on a fibre F of the first projection, and other 4, P_1', P_2', P_3', P_4' on a fibre $F' \neq F$ of the same projection. The branch locus of the projection $V \to Z$ equals the union of the proper inverse transforms onto Z of the curves F, F', and four fibres N_1, N_2, N_3, N_4 of the second projection, each of them N_i passing through P_i and P_i'. These N_i correspond to the sections S_i on V, L_0, L_0' correspond to the curves E_0, E_0', and the lines blown up from the points P_i, P_i' correspond to the curves E_i, E_i'. Consider the rational map $Z \to \mathbb{P}^2$ which is the composition of the blowing down $Z \to \mathbb{P}^1 \times \mathbb{P}^1$ and the linear projection of the quadric onto $\mathbb{P}^2$ with center at some point lying outside the branch locus of $V \to Z \to \mathbb{P}^1 \times \mathbb{P}^1$. Then the image of the branch locus will be equal to the union of six lines, two of them passing through some point A_1, say ℓ_0, ℓ_0', and four of them passing through other point $A_2 \neq A_1$, say n_1, n_2, n_3, n_4. The pencil of elliptic curves on V is obtained from the pencil of lines through A_1, the curves F_1 and F_2 correspond to some lines m_1 and m_2 through A_1. Let $B_1 = m_1 \cap \ell_3$, $B_2 = m_2 \cap \ell_4$. The pencil P on V corresponds to the pencil of conics passing through A_1, A_2, B_1 and B_2. To get the case B) we just take for D_1 a conic from this pencil passing through the point $\ell_0 \cap n_1$; in the case C) we take D_1 as in B), and for D_2 take a conic from this pencil passing through the point $\ell_0' \cap n_1$. To get the case D) we take for D_1 a conic from the pencil passing through the points $\ell_0 \cap n_1$ and $\ell_0' \cap n_2$ (that can be done only for some special choice of the lines), and D_2 as in C). Finally, to get the case E) we take for D_1 the same conic as in D), and for D_2 the conic passing through the points $\ell_0 \cap n_2$ and $\ell'_0 \cap n_1$ (also take some special choice of the lines) .

Now we will compute the torsion of Burniat's surfaces. Obviously, we have the following torsion divisors of order 2:

Case A): $g_1 = \bar{E}_0 - \bar{E}'_0$, $g_2 = \bar{E}_0 - \hat{F}_1 - \bar{R}_1$, $g_3 = \bar{E}_0 - \hat{F}_2 - \bar{R}_2$,

$g_4 = \bar{S}_2 - \bar{S}_4 - \bar{R}_1 g_5 = \bar{S}_2 - \bar{S}_3 - \bar{R}_2$, $g_6 = \hat{D}_1 - \hat{F}_1 - \bar{S}_3$,

Case B): $g_1 = \bar{E}'_0 - \hat{F}_1 - \bar{R}_1$, $g_2 = \bar{E}'_0 - \hat{F}_2 - \bar{R}_2$, $g_3 = \bar{S}_2 - \bar{S}_4 - R_1$,

$g_4 = \bar{S}_2 - \bar{S}_3 - \bar{R}_2$, $g_5 = \hat{D}_2 - \hat{F}_1 - \bar{S}_3$

Case C): $g_1 = \hat{F}_1 + \bar{R}_1 - \hat{F}_2 - \bar{R}_2$, $g_2 = \bar{S}_2 - \bar{S}_3 - R_2$,

$g_3 = \bar{S}_2 - \bar{S}_4 - \bar{R}_1$, $g_4 = \hat{D}'_2 + R_1 - \bar{E}'_0 - \bar{S}_3$,

Case D): $g_1 = \hat{F}_1 + \bar{R}_1 - \hat{F}_2 - \bar{R}_2$, $g_2 = S_3 + R_2 - S_4 - R_1$,

$g_3 = \hat{D}'_1 + R_1 + R_2 - \bar{S}_2 - \bar{E}_0$.

Case E): $g_1 = \hat{F}_1 + \bar{R}_1 - \hat{F}_2 - \bar{R}_2$, $g_2 = \bar{S}_3 + \bar{R}_2 - \bar{S}_4 - \bar{R}_1$.

(where $\sigma_*(r^*(p^{-1}(D'_2))) = 2\hat{D}'_2$ and $\bar{S}_i = \sigma_*(r^*(p^{-1}(S_i)))$, $i=3, 4$) .

We will show that, in fact, these divisors generate the whole torsion group.

<u>Lemma</u>. Let ${}_2\mathrm{Tors}(X)$ denote the subgroup of elements of order 2 in $\mathrm{Tors}(X)$. Then

$$\mathrm{Tors}(X) = {}_2\mathrm{Tors}(X) .$$

<u>Proof</u>. Let $\delta: X \to X$ be the involution of the second order induced by the rational double projection of X onto V . Then δ induces an automorphism of Tors(X) of order 2 $\delta^*: \mathrm{Tors}(X) \to \mathrm{Tors}(X)$.

For any $g \in \mathrm{Tors}(X)$ the divisor $g+\delta^*(g)$ is invariant with respect to δ and hence being taken twicely comes from a torsion divisor on V. Since V is rational, we get that the latter is linearly equivalently to zero. Thus

$$2(g + \delta^*(g)) = 2g + \delta^*(2g) \sim 0 .$$

Replacing g by 2g we get that $\mathrm{Tors}(X) \neq {}_2\mathrm{Tors}(X)$ implies the existence of a non-trivial torsion divisor g such that $g + \delta^*(g) \sim 0$.

Let D_g be an effective divisor from the linear system $|K_X + g|$, where g as above . Then

$$D_g + \delta^*(D_g) \sim D_g + D_{\delta^*(g)} \sim D_g + D_{-g} \in |2K_X|$$

Using the computation of $2K_X$ on the page 90 we get that there exists a curve

$$C \in |F_o + F_o' + 4S_1 + 2E_1 + 2E_1' + E|$$

(E is a linear combination of other E_i ,E_i') such that

$$D_g + \delta^*(D_g) = \sigma_*(r^*(p^*(C))) .$$

Since $p^{*}(C)$ splits under the covering $r : X' \to V'$, it must touch the branch curve $W' = p^{-1}(W) + R'_1 + R'_2$. Counting the intersection indices we easily find that $p^{*}(C)$ touches the curves $p^{-1}(F_1)$ and $p^{-1}(F_2)$ at one point P_1 and P_2 respectively, and touches the curves $p^{-1}(D_1)$ (or $p^{-1}(D'_1)$ and $p^{-1}(D_2)$ (or $p^{-1}(D'_2)$) at two points P_3, P'_3 and P_4, P'_4 respectively. Also, it does not touch the components E_i or E'_i of W'.

Now notice that both W' and $p^{*}(C)$ are invariant with respect to the automorphism h of V' induced by the inversion automorphism of the elliptic pencil. This shows that the points P_1 and P_2 are fixed under h (and hence are situated on one of the sections $p^{-1}(S_i)$), and the points P_3 and P'_3 (resp. P_4, P'_4) are conjugate with respect to h. Using this we observe that any curve $C' \sim p^{*}(C)$ which passes through P_1 and P_2 and touches $p^{*}(C)$ at P_3 and P_4 will necessarily touch $p^{*}(C)$ at all 6 points $P_1, P_2, P_3, P'_3, P_4, P'_4$. Since $\dim |p^{*}(C)| = \dim |2K_X| = 6$ we always can choose such C'. Considering $r^{*}(C')$ we get the contradiction in view of the following:

<u>Sublemma</u>. Let F be a non-singular projective surface with $q(F) = 0$, D_1 and D_2 effective divisors such that $D_1 - D_2$ is a non-trivial torsion divisor. Then for any D $|D_1+D_2|$ with no common component with $D_1 + D_2$ there exists a point $P \in F$ such that $(D \cdot D_1)_P \neq (D \cdot D_2)_P$.

<u>Proof</u>. Assume the contrary, let $D \in |D_1 + D_2|$ which does not satisfy the assertion of the lemma. Consider the linear pencil generated by the divisors D and $D_1 + D_2$. Resolving its base points we get a morphism $f : F' \to \mathbb{P}^1$ of a surface F' birationally equivalent to F onto $\mathbb{P}^1$ with a fibre containing two numerically equivalent components.

The main lemma of Chapter 2, §1 shows that it is possible only in the case when the general fibre of f is disconnected. Moreover, in this case f has to factor through $f' : F' \to B$, where B is a non-rational curve. This of course, contradicts the assumption $q(F) = 0$.

Theorem. Let X be a Burniat surface of linear genus $p^{(1)}$. Then

$$\mathrm{Tors}(X) = (\mathbb{Z}/2)^{p^{(1)}-1} .$$

Proof. We already know that $\mathrm{Tors}(X) = {}_2\mathrm{Tors}(X)$ and, even more, that any torsion divisor class is invariant with respect to the involution induced by the projection $r : X' \to V'$. Consider the morphism $f : X \to \mathbb{P}^1$ which is defined by the inverse image of the elliptic pencil on V'. We have the following multiple fibres of this morphism:

Case A): $2\bar{E}_0,\ 2\bar{E}_0',\ 2\hat{F}_1 + 2R_1,\ 2\hat{F}_2 + 2R_2$;

Case B): $2\bar{E}',\ 2\hat{F}_1 + 2R_1,\ 2\hat{F}_2 + 2R_2$;

Case C), D), E): $2\hat{F}_1 + 2R_1,\ 2\hat{F}_2 + 2R_2$.

Let $\mathrm{Tors}_f(X)$ be the subgroup of $\mathrm{Tors}(F)$ generated by components of fibres of f. Using the main lemma from Chapter 2, §1 we see that

$$\mathrm{Tors}_f(X) = \begin{cases} (\mathbb{Z}/2\mathbb{Z})^3 & \text{in case A)} \\ (\mathbb{Z}/2\mathbb{Z})^2 & \text{in case B)} \\ \mathbb{Z}/2\mathbb{Z} & \text{in cases C), D), E)} \end{cases}$$

and can be generated by the first three (resp. two, resp. one) divisors g_i indicated on page 94.

Let X_η be the general fibre of f. The restriction homomorphism $\mathrm{Pic}(X) \to \mathrm{Pic}(X_\eta)$ induces the imbedding

$$\mathrm{Tors}(X)/\mathrm{Tors}_f(X) \hookrightarrow {}_2\mathrm{Pic}(X_\eta)^r ,$$

where $\mathrm{Pic}(X_\eta)^r$ denotes the subgroup of divisors on X which are invariant with respect to the automorphism induced by the projection $r_\eta : X_\eta \to V_\eta$, V_η being the general elliptic fibre on V. The covering r_η is ramified along the two points defined by the curves D_1 (or D_1') and D_2 (or D_2').

This shows that each $D \in \mathrm{Pic}(X_\eta)^r$ can be represented by a linear combination of the curves $\hat{D}_1, \hat{D}_2, \bar{S}_1, \bar{S}_2, \bar{S}_3, \bar{S}_4$ (the latter four generates $\mathrm{Pic}(V_\eta)$). Using the relations on V

$$2S_i \sim 2S_j \quad \text{modulo} \quad E_i, E_i'$$

$$D_i \sim 2S_j \quad \text{modulo} \quad E_i, E_i'$$

we find that each divisor $\hat{D}_i - \bar{S}_j$, $\bar{S}_i - \bar{S}_j$ defines an element of ${}_2\mathrm{Pic}(X_\eta)^r$,

Now we notice that the covering $r : X' \to V'$ is defined by the line bundle corresponding to the divisor

$$p^*(3F+2S_1) - R_1' - R_2' - 3R_1 - 3R_2 \quad \text{mod.} \quad E_i, E_j'$$

(see p. 88). This implies that

$$\hat{D}_1 + \hat{D}_2 \sim 2\bar{S}_2 \quad \text{modulo components of fibres of } f .$$

There is also a relation between $\bar{S}_i$

$$\overline{S}_1 + \overline{S}_2 \sim \overline{S}_3 + \overline{S}_4 \quad \text{modulo components of fibres of } f$$

because S_i defines the 4 points of order 2 on V.

Summarizing we get that ${}_2\mathrm{Pic}(X_\eta)^r$ is generated by the three divisors

$$\overline{S}_3 - \overline{S}_2 \,, \quad \overline{S}_4 - \overline{S}_2 \,, \text{ and } \hat{D}_1 - \overline{S}_3$$

which as it is easily checked are independent.

The arguments above show that any element of $\mathrm{Tors}(X)/\mathrm{Tors}_f(X)$ can be represented by a sum of the above divisors plus a combination of components of fibres of f. It is easy to find in each of the cases A)-E) the corresponding torsion divisors. In fact, we obtain that these divisors are combinations of divisors g_i ($i=4, 5, 6$ in case A), $i=3, 4, 5$ in case B), $i=2, 3, 4$ in case C), $i=2, 3$ in Case D), $i=2$ in case E)) indicated on p. 73. This proves the theorem.

Remark. As we observed above the morphism $f : X \to \mathbb{P}^1$ has 4 multiple fibres of multiplicity 2 in case A). Let $B \to \mathbb{P}^1$ be the 2-sheeted covering of $\mathbb{P}^1$ by an elliptic curve B branched at the four points corresponding to the multiple fibres. The normalization X' of the surface $X \times_{\mathbb{P}^1} B$ is a double covering of X non-ramified outside the two points $Q_1 = \hat{F}_1 \cap R_1$ and $Q_2 = \hat{F}_2 \cap R_2$. Also, $\overline{X}$ being mapped onto B has the infinite fundamental group, the points $\overline{Q}_1$ and $\overline{Q}_2$ lying over Q_1 and Q_2 are ordinary double points. This shows that the complement $X - \{Q_1, Q_2\}$ has a non-ramified covering with infinite fundamental group, hence X itself has infinite fundamental group.

Another way to prove that the fundamental group of the Burniat surface with $p^{(1)} = 7$ is infinite is based on the corollary to Lemma 1 of Chapter III, §1 . Consider the surface X_T corresponding to the torsion group T of X . Then we have

$$g(X_T) = \sum_{g \in T} h^1(g) = \sum_{\substack{g \in T \\ g \neq 0}} (h^0(K_X + g) - 1)$$

Consider the inverse image of the pencil $\mathcal{P}$ onto X . The divisor $2\hat{D}_1$ belongs to this pencil and $h^0(2\hat{D}_1 + \overline{R}_1 + \overline{R}_2) = 2$. Now

$$2(\overline{R}_1 + \overline{R}_2 + 2\hat{D}_1 - K_X) \sim 2(2\hat{D}_1 - 2\overline{F} - 2\overline{S}_1 + \overline{E}_0 + \overline{E}_0' + 2\overline{R}_1 + 2\overline{R}_2)$$

$$2(\overline{F} + 2\overline{S}_1 - 2\overline{R}_1 - 2\overline{R}_2 - 2\overline{F} - 2\overline{S}_1 + \overline{E}_0 + \overline{E}_0' + 2\overline{R}_1 + 2\overline{R}_2) \sim 2(\overline{E}_0' - \overline{E}_0) \sim 0 .$$

This shows that $2\hat{D}_1 + \overline{R}_1 + \overline{R}_2 \sim K_X + g$ and hence

$$q(X_T) > 0 .$$

This, of course, implies that X_T and thus X has infinite fundamental group.*)

*) See Epilogue

§5. Surfaces with $p^{(1)} = 9$.

Such surfaces were constructed by M. Kuga [29] and A. Beauville [3].

Kuga's construction:

Let $H = \{z \in \mathbb{C} : \mathrm{Im}(z) > 0\}$ be the upper half plane. The Lie group $\mathbb{P}GL(2,\mathbb{R}) = \mathbb{SL}(2,\mathbb{R})/\pm 1$ is identified in a natural way with its group of analytic automorphisms.

Let Γ be a discrete subgroup of $\mathbb{P}GL(2,\mathbb{R}) \times \mathbb{P}GL(2,\mathbb{R})$ acting freely on $H \times H$ with compact quotient $V = H \times H/\Gamma$. By Matsushima-Shimura [31] we have

$$h^{1,0}(V) = h^{0,1}(V) = q(V) = 0 ;$$

$$h^{1,1}(V) = 2p_g(V) + 2 .$$

Therefore,

$$c_2(V) = 4p_g(V) + 4 , \quad K_V^2 = 8p_g(V) + 8 .$$

Next, notice that V has no exceptional curves of the first kind (and more generally, no rational curves), because the projection $H \times H \to H \times H/\Gamma = V$ splits over such curve, but $H \times H$ does not contain any complete curves.

Thus, to find the needed surfaces with $p_g(V) = 0$ and $p^{(1)} = 9$ it suffices to choose such Γ that

$$c_2(H \times H/\Gamma) = 4 .$$

By the Gauss-Bonnet formula

$$c_2(V) = \frac{1}{4\pi^2} \operatorname{vol}(V) \quad ,$$

where the volume $\operatorname{vol}(V)$ is computed by integration of the invariant volume element

$$d\nu = \frac{dx_1 \wedge dy_1}{y_1^2} \wedge \frac{dx_2 \wedge dy_2}{y_2^2} \quad ,$$

($(z_1, z_2) = (x_1+iy_1, x_2+iy_2)$ being the coordinates on $H \times H$) .

Now, let

$k = \mathbb{Q}(\sqrt{d})$ be a real quadratic field, d the discriminant;

$A = A(k,\theta)$ be the division quaternion algebra with the center k and with the discriminant $\theta = p_1 p_2 \cdots p_{2r}$ assumed to be totally indefinite (that is, $A \otimes_{\mathbb{Q}} \mathbb{R} = M_2(R) \oplus M_2(R)$)

$N : A \to k$ be the reduced norm of A ;

$\underline{O}$ be the maximal order of A (unique up to conjugation if the class number of k equals 1);

$E(\underline{O})$ be the group of all units of $\underline{O}$;

$\overline{\Gamma} = \{g \in E(\underline{O}) : N(g) = 1\}$

Consider the natural injection $i : A \to A \otimes_{\mathbb{Q}} \mathbb{R} = M_2(\mathbb{R}) \oplus M_2(\mathbb{R})$ and the projection $j : GL_2(\mathbb{R}) \times GL_2(\mathbb{R}) \to \mathbb{P}GL(2,\mathbb{R}) \times \mathbb{P}GL(2,\mathbb{R})$. Let $\Gamma = j(i(\overline{\Gamma}))$ be a discrete subgroup of $\mathbb{P}GL(2,\mathbb{R}) \times \mathbb{P}GL(2,\mathbb{R})$ with compact quotient $V = H \times H/\Gamma$; we note that Γ is isomorphic to the image of $\overline{\Gamma}$ into A^*/k^* .

According to Simizu ([42]) the volume $\operatorname{vol}(H \times H/\Gamma)$ can be expressed through the zeta function $\zeta_k(s)$ of k by the formula:

$$\mathrm{vol}(H \times H/\Gamma) = \frac{2}{\pi^2} d^{3/2} \zeta_k(2) \Pi(|p_i| - 1)$$

($|p|$ denotes the norm of prime ideal p of k) .

Now

$$\zeta_k(s) = \zeta(s) L(s,\chi) ,$$

where $\zeta(s)$ is the Riemann zeta function and $L(s,\chi)$ is the Dirichlet L-function associated with the character χ mod d

$$\chi(n) = \begin{cases} (\frac{n}{d}) & ; \text{ if } d \equiv 1 \mod 4 \\ (\frac{n}{m})(-1)^{(n-1)/2} & , \text{ if } d = 4m, m \equiv 3 \mod 4 \\ (\frac{n}{m'})(-1)^{(n^2-1)/8} & , \text{ if } d = 8m', m \equiv 1 \mod 4 \\ (\frac{n}{m'})(-1)^{(n^2-1)/8+(n-1)/2} & , \text{ if } d = 8m', m' \equiv 3 \mod 4 \end{cases}$$

The value of the Riemann zeta at 2 equals $\pi^2/6$. The value $L(s,\chi)$ at 2 equals

$$L(2,\chi) = \frac{1}{2}(\frac{2}{d})^2 \tau(\chi) B_d ,$$

where

$$\tau(\chi) = \sum_{n=1}^{d-1} \chi(n) e^{2\pi i n/d} \quad , \text{ the Gauss' sum}$$

$$B_d = \frac{1}{2d}(\sum_{m=1}^{d-1} m^2 \chi(m)) .$$

Thus, we have

$$c_2(H \times H/\Gamma) = \frac{2d^{3/2}}{2^2\pi^4} (\frac{\pi^2}{6}) \frac{1}{2} (\frac{2\pi}{d})^2 \tau(\chi) B_d \Pi_{p/\theta}(|p|-1) =$$

$$= \frac{1}{6} \frac{1}{d^{1/2}} \tau(\chi) B_d \Pi(|p|-1) .$$

Since the Gauss' sum $\tau(\chi)$ has absolute value $|\tau(\chi)| = d^{1/2}$ and $c_2 = \frac{1}{4\pi^2}$ vol is positive, we get

$$c_2(HxH/\Gamma) = \frac{1}{6}|B_d| \prod_{p|\theta}(|p| - 1)$$

Next, we have to be assured that the group Γ acts freely on HxH, and hence HxH/Γ is smooth. Since the stabilizator group of any point is a finite subgroup of Γ, that can be if and only if Γ has no elements of finite order.

Let $g \in \Gamma$ be an element of order N, $\bar{g} \in \bar{\Gamma}$ some of its preimages in $\bar{\Gamma}$. We have $g^N = \pm 1$, and thus $\bar{g}^{2N} = 1$. Then the quaternion algebra A has to contain a subfield isomorphic to the field

$$\mathbb{Q}(e^{\pi i/N}) = \mathbb{Q}(\bar{g}) .$$

Conversely, if the class number $h(k) = 1$, then $A \supset \mathbb{Q}(e^{2\pi i/N})$ implies that $\bar{\Gamma}$ has an element of order N.

Since the maximal subfield of A has degree 2 over k, we have

$$\phi(N) = [\mathbb{Q}(e^{2\pi i}) : \mathbb{Q}] \qquad \text{divides } 4 .$$

Thus the only possible orders for N are

$$N = 2,3,4,5,6,8,10,12 .$$

Obviously, an element of order 2 in $\bar{\Gamma}$ defines the unit element of Γ

Now, if $\phi(N) = 2$ (N=3,4,6) then the maximal subfield K of A coincides with $k(e^{2\pi i/N})$, if $\phi(N)=4$ then $K = \mathbb{Q}(e^{2\pi i/N})$ and k is the real quadratic subfield of $\mathbb{Q}(e^{2\pi i/N})$.

Let K be a quadratic extension field of k; then the local arguments show that K is embedable into $A = A(k,\theta)$ if and only if $p|\theta$ does not decompose in K.

Now we are ready to give an explicit example.

Example. $k = Q(\sqrt{2})$, $d = 8$, $\theta = p_2p_5$, where p_2 and p_5 lie over 2 and 5 accordingly.

We compute

$$B_8 = 1 \ , \ c_2(H\times H/\Gamma) = \frac{1}{6} B_8(2-1)(25-1) = 4 \ .$$

To check the smoothness of $H\times H/\Gamma$ we observe that the only cyclotomic field containig k is $\mathbb{Q}(e^{2\pi i/8})$, and in this case p_2 and p_5 do not decompose. Thus it suffices to consider the cases $N = 3,4,$and 6. In the second case $K = \mathbb{Q}(\sqrt{2},i)$, and in the first and the third, $K = \mathbb{Q}(\sqrt{2},\sqrt{-3})$. In the both cases we easily verify that p_2 and p_5 do not decompose.

Notice that other examples can be also obtained by taking instead of some other discrete subgroups in $\underline{O}$, for example,

$$\bar{\Gamma}' = \{g \in E(\underline{O}) : N(g) \text{ is a totally positive unit of } k\} \ .$$

We refer to [29] for the examples of the corresponding surfaces $H\times H/\Gamma$.

To compute the torsion group $\mathrm{Tors}(H\times H/\Gamma)$ we note that

$$\mathrm{Tors}(H^2(V,Z)) = H_1(V,Z) = \Gamma/[\Gamma,\Gamma] = \bar{\Gamma}/(\underline{+}1)[\bar{\Gamma},\bar{\Gamma}] \ .$$

For any maximal two-sided ideal $p\underline{O}$ in $\underline{O}$ we may consider the image $\phi(\bar{\Gamma})$ in $\underline{O}/p$ ($= M_2(\mathbb{F}_q)$ or $\mathbb{F}_{q^2}$, $q=\mathrm{Norm}_{k/\mathbb{Q}}(p)$, depending on whether $p\nmid\theta$ or $p|\theta$).

Moreover, by the Eichler approximation theorem we have

$$\phi(\overline{\Gamma}) = \begin{cases} SL_2(\mathbb{F}_q) & , p \nmid \theta \\ U & , p \mid \theta \end{cases}$$

where $U = \{ a \in \mathbb{F}_{q^2} : N_{\mathbb{F}_{q^2}/\mathbb{F}_q}(a) = 1 \}$ is a cyclic group of order $q+1$.

This immediately shows that it is always

$$\mathrm{Tors}(V) = \overline{\Gamma} / (\pm 1)[\overline{\Gamma}, \overline{\Gamma}] \neq 1 \quad .$$

The more detailed analysis gives the following result:

Theorem([29]). There exists a subgroup M of Γ containing $[\Gamma,\Gamma]$ such that

$$\overline{\Gamma}/M = \bigoplus_{i=1}^{r} \mathbb{Z}/(q_i+1) \oplus (\mathbb{Z}/2)^a \oplus (\mathbb{Z}/3)^b \quad ,$$

where

$$q_i = N_{k/\mathbb{Q}}(p_i) \ , \ \theta = p_1 \cdots p_r \ ;$$

$$a = \begin{cases} 2 & \text{if } (2) = p_2p_2' \ , \ p_2 \neq p_2' \text{ and } p_2 \nmid \theta \ , p_2' \nmid \theta \\ 1 & \text{if } p_2 \mid 2 \ , \ |p_2| = 2 \ , \ p_2 \nmid \theta \quad \text{but other divisor of 2 divides } \theta\theta \\ 0 & \text{otherwise} \end{cases}$$

$$b = \begin{cases} 2 & \text{if } (3) = p_3p_3' \ , \ p_3 \neq p_3' \text{ and } p_3 \nmid \theta \ , \ p_3' \nmid \theta \\ 1 & \text{if } (3) = p_3^2 \ , \ p_3 \nmid \theta \quad \text{or} \quad p_3 \mid 3 \text{ and other divisor of 3 divides } \theta \\ 0 & \text{otherwise} \end{cases}$$

Moreover, $M = [\overline{\Gamma}, \overline{\Gamma}]$ if the congruence subgroup conjecture of Bass-Serre is true for $\overline{\Gamma}$. Also, $-1 \in M$ if and only if one of $q_i \equiv 1 \mod 4$.

In the above example we have

$$\overline{\Gamma}/M = \mathbb{Z}/3 \oplus \mathbb{Z}/6 .$$

Beauville's examples ([3]).These surfaces are constructed as the quotients $V = C\times D/G$, where C and D are complete non-singular algebraic curves of genus g at least 2, G is a finite group acting freely on the product.

To construct the quotient with the needed properties Beauville proposes to take for G a finite group of order $(g(C)-1)(g(D)-1)$ acting on the both C and D with the rational quotients. In order to get a free action on $C\times D$ he puts

$$g(x,y) = (g(x),\sigma(g)(y)) \quad , \quad g\in G \quad , \quad (x,y)\in C\times D \quad ,$$

where σ is án automorphism of G such that for all $g\in G$ acting non-freely on C $\sigma(g)$ acts freely on D .

In virtue of the lemma of Chap.I,§2 we have

$$p_a(V) = 0 \quad , \quad K_V^2 = 8 \quad .$$

Moreover, V does not contain any rational curves, since the projection $C\times D \to V$ has to split over such curve and there are no rational curves on $C\times D$. This implies that V is a minimal model.

It remains to prove that the irregularity $q(V) = 0$. We have

$$H^1(V,\mathcal{O}_V) = H^1(C\times D,\mathcal{O}_{C\times D})^G = H^1(C,\mathcal{O}_C)^G \oplus H^1(D,\mathcal{O}_D)^{\sigma(G)}$$

but, since C/G and $D/\sigma(G)$ are rational curves, the both summands are zeros.

Example 1. $C = D$ is the plane curve with the equation:

$$x^5 + y^5 + z^5 = 0 ,$$

$G = (Z/5)^2$ acts on C by the formulas:

$$(p,q)(x,y,z) = (\xi^p x, \xi^q y, z) , \quad \xi = e^{2\pi i/5} ,$$

σ is the automorphism of G given by $(1,0) \to (1,1), (0,1) \to (1,2)$.

The set of elements of G which act freely is $A = \{(p,q), p \neq q\}$ and $G = \{1\} \cup A \cup \sigma(A)$.

Example 2. $C = D$ is the curve of genus 4 given by the equation in $\mathbb{P}^3$:

$$x^3 + y^3 + z^3 + t^3 = 0 , \quad xy + zt = 0 .$$

$G = (Z/3)^2$ acts on C by the formulas:

$$(p,q)(x,y,z,t) = (\xi^p x, \xi^{-p} y, \xi^q z, \xi^{-q} t) , \quad \xi = e^{2\pi i/3} ,$$

σ is the automorphism given by $(1,0) \to (1,1)$, $(0,1) \to (1,2)$.

The set of elements of G acting freely on C is the set $A = \{(p,q), p \pm q \neq 0\}$ and $G = \{1\} \cup A \cup \sigma(A)$

Applying the well known Hochshild-Serre exact sequence:

$$0 \to \mathrm{Hom}(G, \mathbb{C}^*) \to \mathrm{Pic}(C \times D/G) \to \mathrm{Pic}(C \times D)^G \to H^2(G, \mathbb{C}^*)$$

we see that

$$\mathrm{Tors}(C \times D/G) \supset G/[G,G] .$$

In particular, in the above examples the torsion group is non-trivial.

6. Concluding remarks.

It would be very optimistic to expect the complete classification of all surfaces of general type with $p_g=0$. However, there are still many problems to answer in the visible future.

One of the most interesting from my point of view is the following:

Problem 1. Is there a simply connected surface of general type with $p_g=0$?

Or more weak

Problem 1'. Is there a surface of general type with $p_g=0$ and trivial torsion group?

Consider the class of all surfaces of general type with $p_g=0$ and fixed $P_2= p^{(1)}$. Then there exists a number N such that the N-canonicla system defines a birational morphism for all such surfaces([4]). Thus the set of its N-canonicla models can be parametrized by an open subset of the Hilbert scheme corresponding to some Hilbert polynomial. Since the latter is of finite type, this open subset consists of finite number of connected components. The surfaces parametrized by a connected variety are diffeomorphic, and,in particular,have the same fundamental group. This argument shows that there are only finite number of possibilities for the fundamental group of a surface. In particular, the order of the torsion group is bounded by a constant depending only on $p^{(1)}$.

Problem 2. Find a bound for the order of the torsion group of surfaces with the fixed $p^{(1)}$ (as always of general type and with $p_g=0$).

We remind that it is done in the cases of numerical Godeaux and Campedelli surfaces.

Consider the class of all surfaces with the fixed value

the torsion group T. Denote it by $\mathcal{M}$ (a,T).

Problem 3. Can $\mathcal{M}$(a,T) be parametrized by a connected variety? In particular, are the elements of $\mathcal{M}$(a,T) diffeomorphic to each other?

For the start it would be very interesting to know the answer at least in the cases $\mathcal{M}$(2,Z/2), $\mathcal{M}$(2,Z/3) and $\mathcal{M}$(3,Z/2⊕Z/2) . Recall that in the last case we know two (and possibly even three) different constructions of surfaces from this class. In some cases the answer is positive (e.g. $\mathcal{M}$(2,Z/4), $\mathcal{M}$(2,Z/5), $\mathcal{M}$(3,abelian of order 8)).

We still do not know if all possible values of $p^{(1)}$ are realized*).

Problem 4. Are there surfaces with $p^{(1)} = 8$ and 10 ?

There is much hope to solve the following

Problem 5. Find all possible torsion groups of numerical Godeaux and Campedelli surfaces.*)

The validity of the following assertion is observed in all known examples:

Problem 6. Prove that the fundamental group is infinite in the case $p^{(1)} \geq 7$ and finite otherwise.

---

*) See Epilogue

BIBLIOGRAPHY

[1] Algebraic surfaces, ed.I.R.Shafarevich. Proc.Steklov Math.Inst.,vol.75, 1965(Engl.transl., AMS,1967).

[2] Artin M., On Enriques' surface. Harvard thesis,1960.

[3] Beauville A., A letter to the author of September 25,1977.

[4] Bombieri E., Canonical models of surfaces of general type,Publ.Math.I.H.E.S., 42(1973),171-219.

[5] ________, Catanese F., A non-published manuscript.

[6] ________, Mumford D., Enriques' classification of surfaces in char.p.II, "Complex Analysis and Alg.Geometry",Cambridge Univ.Press,1977,p.23-42.

[7] Burniat P., Sur les surfaces de genre $P_{12}>0$, Ann.Math.Pura et Appl.(4), 71(1966),1-24.

[8] ________, Surfaces algébriques régulières de genre géométrique $p_g=0,1,2,3$ et de genre lineaire $3,4,\dots,8p_g+7$. Colloque de Géom.Algébr. du C.B.R.M.,Bruxelles,1959,129-146.

[9] Campedelli L.,Sopra alcuni piani doppi notevoli con curva di diramazione del decimo ordine. Atti Acad.Naz.Lincei,15(1932),358-362.

[10] Castelnuovo G., Sulle superficie di genere zero. Memorie Soc.Ital.Scienze, (3),IO (1896),103-123.

[11] Catanese F., Pluricanonical mappings of surfaces with $K^2=1,2$ and $q=p_g=0$, these proceedings, pp.

[12] Cohomologie etale (SGA $4\frac{1}{2}$), ed.P.Deligne, Lect.Notes Math., vol.569,1977 .

[13] Dolgachev I., On rational surfaces with a pencil of elliptic curves. Izv.Akad.Nauk SSSR(ser.math.),30(1966),1073-1100(in russian).

[14] ________, On Severi's conjecture on simply connected algebraic surfaces. Dokl.Akad,Nauk SSSR,170(1966),249-252(Sov.Math. Doklady,7(1966),1169-1172).

[15] ________, Weighted projective varieties (preprint).

[16] Enriques F., Le superficie algebriche. Bologna.1949.

[17] ________, Un' osservazione relativa alle superficie di bigenere uno. Rendiconti Acad.Sci.Bologna, 1908,40-45.

[18] Godeaux L., Sur une surface algébrique de genere zero et de bigenre deux. Atti Acad.Naz.Lincei,14(1931),479-481.

[19] ________, Les surfaces algébriques non rationnelles de genres arithmetique et géométrique nuls (Actualites scient.,No 1230, Paris,1933.

[20] ________, Sur la construction de surfaces non rationnelles de genres zero. Bull.Acad.Royale de Belgique,45 (1949),688-693.

[21] ________, Les surfaces algébriques de genres nuls a courbes bicanoniques irreductibles. Rendiconti Circolo Mat.Palermo,7(1958),309-322.

[22] ________, Recherches sur les surfaces non rationnelles de genres arithmetiques et géométriques nuls.Journ.Math.Pures et Appl., 44 (1965),25-41.

[23] ________, Surfaces non rationnelles de genres zero, Bull.Inst.Mat.Acad. Bulgare,12(1970),45-58.

[24] Grothendieck A., Le groupe de Brauer, II ,III, "Dix exposes sur la cohomologie des schémas",Amsterdam-Paris,North-Holland, 1968,pp.66-188.

[25] Iithaka S., Deformations of compact complex surfaces,III. Journ.Math. Soc.Japan,23(1971),692-705.

[26] Kodaira K., On compact analytic surfaces,II. Ann.Math.,77(1963),563-626.

[27] ________ , On the structure of compact complex analytic surfaces,I. Amer.Journ.Math.,86(1964),751-798.

[28] ________ , On homotopy K3-surfaces, "Essays on Topology and related topics",Springer,1970,pp.58-69.

[29] Kuga M., preprint.

[30] Magnus W., Noneuclidean tesselations.Acad.Press.1974.

[31] Matsushima Y.,Shimura G., On the cohomology groups attached to certain vector valued differential forms on the product of the upper half planes.Ann.Math., 78(1963),417-449.

[32] Miyaoke Y., Tricanonical maps of numerical Godeaux surfaces. Inv.Math., 34(1976),99-111.

[32]' ________ , On numerical Campedelli surfaces."Complex analysis and Algebraic geometry",Cambridge Univ.Press,1977,pp.112-118.

[33] Mumford D., Pathologies ,III. Amer.Journ.Math.,89(1967),94-104.

[34] Ogg A., Cohomology of abelian varieties over function fields. Ann.Math.,76(1962),185-212.

[35] Oort F., A letter to C.Peters of February 1976.

[36] Peters C., On two types of surfaces of general type with vanishing geometric genus. Inv.Math.,32(1976),33-47.

[37] ________ , On some Burniat's surfaces with $p_g=0$., preprint

[38] Raynaud M., Caracteristique de Euler-Poinçaré d'un faisceaux constructible et cohomologie des variétes abéliennes. "Dix exposes sur la cohomologie des schémas",North-Holland, 1968,pp.12-31.

[39] Reid M., Some new surfaces with p_g=0, preprint.

[40] ______ , A letter to D.Mumford of December 22,1975.

[41] Shafarevich I., Principal homogeneous spaces defined over a function field. Trudy Steklov Math.Inst.,64(1961),316-346 (Transl.AMS,vol.37,85-115).

[42] Shimizu H., On discontinuous groups operating on the product of the upper half planes.Ann.Math.,77(1963),33-71.

[43] Zarisky O., Algebraic surfaces. Springer. 1974.

EPILOGUE

After this work has been almost done the author was informed in many new results.

1. Numerical Godeaux surfaces with Tors = $\mathbb{Z}/3$ have been constructed by Miles Reid [45]. The construction is very delicate.

2. The final version of Peters' preprint [37] has been published [44].It can be found there the result about the torsion of Burniat's surfaces(the proof is not complete). Also it is proven there that the fundamental group is infinite in case $p^{(1)}=7$. This result is also refered to M.Reid.

3. F.Oort and C.Peters also have proven that the torsion of Campedelli--Oort-Kulikov surfaces with $p^{(1)}=2$ is equal to $\mathbb{Z}/2$ ([51]).

4. M.Inoue has constructed surfaces with $p^{(1)}=8$ and also calculated the fundamental group for Burniat's surfaces ([46]).

5. M.Reid has computed the canonical ring of numerical Godeaux surfaces with Tors=$\mathbb{Z}/2$ ([46]).

6. M.Reid has proven that $\#$ Tors ≤ 9 for numerical Campedelli surfaces. He conjectures that 9 can be replaced by 8 and the surfaces with the torsion group of order 8 are the Godeaix-Reid surfaces .Another conjecture: $\#$ Tors < 30 for surfaces with $p^{(1)}=4$ ([47]).

7. Using the nonarchimedean uniformization theory D.Mumford has construced a surface with $p^{(1)}=10$ ([48]).

8. Many people have discovered independently a surface with Tors = $\mathbb{Z}/5$ and $p^{(1)}=3$ ([46]). As it was explained to me by Fabrizio Catanese it can be

constructed in the foloowing way. Let F be a quintic surface in P^3 which is invariant under an involution of order 5 and posseses 20 ordinary double points. Also assume that there exists a quartic surface B tangent to F along a curve C which passes through these double points and smooth at them. The existence of such surfaces F and B is proven in [49]. Blow up F at these 20 double points to the surface $\bar{F}$, then the sum of the twenty exceptional -2-curves on $\bar{F}$ is linearly equivalent to the strict inverse transform of C taken twicely. Let $\bar{V}$ be the double covering of $\bar{F}$ branched at those curves, V the blowing down of the strict trnasforms of the branch locus. Then it can be easily shown that $K_V^2 = 10$, $p_g(V) = 4$. The $\mathbb{Z}/5$-action on F extends to a free action on V and the quotient defines the needed surface X. By Reid's result (see 6.) we get $\mathrm{Tors}(X) = \mathbb{Z}/5$. Moreover, the surface $\bar{V}$ can be realized as a non-singular compactification of a quotient of the upper half planes by a discrete group of Hilbert's type ([50]), this implies that $\bar{V}$ is simply connected, and hence the fundamental group of X is $\mathbb{Z}/5$.

9. C.Peters conjectures that for any double plane of general type with $p_g=0$ the torsion group consists of elements of order 2 ([44]').

Supplementary bibliography.

[44] Peters C. , On certain examples of surfaces with $p_g=0$ due to Burniat. Nagoya Math.Journ.,66(1977),109-119.

[44]' _______ , A letter to the author of March 9, 1978.

[45] Reid M., Surfaces with $p_g=0, K^2=1$. Journ.TokyoMath.Faculty(to appear).

[46] Reid M.; A letter to the author of January 18, 1978.

[47] ______ , A letter to the author of April 24,1978.

[48] Mumford D., An algebraic surface with K ample, $(K^2)=9, p_g=q=0$ (preprint).

[49] Gallarati D., Ricerche sul contatto di superficie algebriche curve. Memoires des Acad. Royale de Belgique., t.32, f.3,

[50] Vanderge G., Zagier D., Hilbert modular group for field $Q(\sqrt{13})$. Inv.Math. 42(1977), 93-131.

[51] Oort F., Peters C., A Campedelli surface with torsiongroup Z/2. (preprint).

CENTRO INTERNAZIONALE MATEMATICO ESTIVO

(C.I.M.E.)

THE THEORY OF INVARIANTS AND ITS APPLICATIONS TO SOME PROBLEMS IN THE ALGEBRAIC GEOMETRY

F. A. BOGOMOLOV

THE THEORY OF INVARIANTS AND ITS APPLICATIONS TO SOME PROBLEMS IN THE ALGEBRAIC GEOMETRY

F. A. Bogomolov
Steklov Institute

1. Introduction to invariant theory

First I want to recall the classical construction of a locally trivial fibre bundle. In the simplest case, the fibre bundle F_γ with fibre F and base X may be constructed as an associated bundle to any principal fibration X_γ with a group G as a fibre: the group G acts on the fibre F and $F_\gamma = X_\gamma \times_G F$, where G acts on F on the right , and on X_γ on the left. A simple, but important, remark: this construction is twice functorial - it is functorial on the base and functorial on the fibre, in the sense that to any morphism $h : F \to Q$ of G - spaces (such that $f \circ g = g \circ h$ for any $g \in G$) we have the morphism of corresponding fibre bundles $h_\gamma : F_\gamma \to Q_\gamma$. The parameter γ here is a cocycle $\gamma \in H^1 (X, \Gamma_G)$ where Γ_G is some subbundle of the sheaf of functions on X with values in G . These facts are proved directly from the diagram:

$$X_\gamma \times F \xrightarrow{\text{id } h} X_\gamma \times Q$$
$$X_\gamma \times_G G \xrightarrow{h} X_\gamma \times_G Q \quad ;$$

THE THEORY OF INVARIANTS AND ITS APPLICATIONS TO SOME PROBLEMS IN THE ALGEBRAIC GEOMETRY

F. A. Bogomolov
Steklov Institute

1. Introduction to invariant theory

First I want to recall the classical construction of a locally trivial fibre bundle. In the simplest case, the fibre bundle F_γ with fibre F and base X may be constructed as an associated bundle to any principal fibration X_γ with a group G as a fibre: the group G acts on the fibre F and $F_\gamma = X_\gamma \times_G F$, where G acts on F on the right, and on X_γ on the left. A simple, but important, remark: this construction is twice functorial - it is functorial on the base and functorial on the fibre, in the sense that to any morphism $h : F \to Q$ of G-spaces (such that $f \circ g = g \circ h$ for any $g \in G$) we have the morphism of corresponding fibre bundles $h_\gamma : F_\gamma \to Q_\gamma$. The parameter γ here is a cocycle $\gamma \in H^1(X, \Gamma_G)$ where Γ_G is some subbundle of the sheaf of functions on X with values in G. These facts are proved directly from the diagram:

$$X_\gamma \times F \xrightarrow{\text{id } h} X_\gamma \times Q$$
$$X_\gamma \times_G G \xrightarrow{h} X_\gamma \times_G Q \quad ;$$

h is well defined because $(1\times h)\circ g = g\circ(1\times h)$ and h_γ is the orbit space.

Remark. If $F \hookrightarrow Q$ is a morphism of G-invariant subspaces, then there is a subbundle $F_\gamma \hookrightarrow Q_\gamma$.

Now suppose that each object in the previous construction has some additional structure and that these structures are compatible with each other. Then the constructed morphism h will be compatible with the structure on the spaces F_γ and Q_γ.

For example if X_γ, X, F, Q, G are algebraic varieties, G is an algebraic group acting on X_γ, and the projections $p: X_\gamma \longrightarrow X$ and $h: F \longrightarrow Q$ are algebraic, then F_γ and Q_γ have the structure of algebraic varieties and $h_\gamma : F_\gamma \to Q_\gamma$ is an algebraic morphism. Now, we can use this construction in the following situation: suppose that G is a linear algebraic group, F, Q are vector spaces with a linear action of group G, both of finite dimension, and X is a compact algebraic variety (or perharps a complex manifold). Then $\gamma \in H^1(X, \mathcal{O}(G))$ and for every G-morphism $F \to Q$ we get $h_\gamma : F_\gamma \to Q_\gamma$ and we can transpose the section from F_γ to Q_γ using the morphism $h^*_\gamma : H^0(X,F)_\gamma \to H^0(X,Q)_\gamma$. This morphism is non-linear if h is non-linear and usually we cannot define any analogous morphism for the higher cohomology group $H^i(X, F)$.

Example. Consider the vector bundle E^k associated to a cocycle $\gamma \in H^1(X, \mathcal{O}(GL(k)))$. Then any tensor product $(E^k)^{\otimes m}$ of E^k is constructed by mean of the same cocycle γ and representation $\gamma^{\otimes m}: GL(k) \to GL(km)$.

Of course the same is true for any other algebraic representation of $GL(k)$. We have a natural morphism, called the Veronese morphism, $\gamma_k : E^k_x \to (E^k_x)^{\otimes m}$, defined by $y \mapsto y^{\otimes m}$.

It commutes with the action of $GL(k)$ on both spaces and we get the Veronese morphism $\nu_k : E^k \to E^{k\otimes m}$.

Now our problem is a follows. Suppose that we can construct a section of $E^{\otimes m}$ where m is large enough. What can be deduced the properties of E ?

For this we have to find a lot of morphisms from $E_x^{\otimes m}$ into "a simpler complex G-space". These simple G-spaces will be called models and we shall look at invariant theory as a theory of morphisms into models. The classical invariant theory, from such point of view, is a theory of morphisms into one model, the trivial G-space $\mathbb{C}$.

Let us take the group G reductive for beginning. Then we define the three kinds of models:

I) the trivial G-modul $\mathbb{C}$,

II) the homogeneous spaces G/H where H is a reductive subgroup of the group G ,

III) the HV-manifolds.

Let us define an HV-manifold. I recall that an irreducible representation of a group G is defined by a character $\chi(T)$ of a maximal torus T and such two representations C_χ , $C_{\chi'}$ coincide if $\chi = w\chi'$ for any $w \in W(G)$ where $W(G)$ is the Weyl group of G. Then there is a highest vector $x_\chi \in T$ such that $T x_\chi = \chi(T)x$. The closure of the orbit Gx is called an HV-manifold, and will be denoted A_χ.

We will refer to the set of A_χ as the set of the models of third type. The A_χ could be described in another way. Take any parabolic subgroup $P \hookrightarrow G$ and the homogeneous one dimensional vector bundle E_χ on the space G/P constructed by mean of a character $\chi(P)$. Now suppose that we can blow-down the zero-section of E to a point, i.e. E is negative. The manifold that we get after this blow-down is an affine HV-manifold and all HV-manifold can be obtained in this way.

Recall the main results of the invariant theory (i.e. the theory of morphisms into the first model, in our definition). All points in affine spaces divide in three types:

a) Stable points: they have the maximal dimensional closed orbit natural projection $p: G \otimes_H W_x \to G/H$, where W_x is the submanifold of H_x-unstable points in N_x. But $f(W_x)$ coincides with the submanifold $W_{G/H}$ and f is a finite morphism from W_x to $W_{G/H}$. Looking at the orbits in W_x and $W_{G/H}$, we can conclude that if $f(x) = f(y)$ then $p(x) = p(y)$ and so p induces a morphism $p: W_{G/H} \to G/H$.

Remark. This is a holomorphic morphism and it can be proved that it is regular, but we shall not need such details for our purposes.

2. Unstability of bundle, sufficient conditions.

We need the following theorem for our program of constructing G-morphisms into the model spaces.

Theorem 3 (Bogomolov). *For any G-invariant affine variety X lying in the manifold of unstable points NS_G there exists a non trivial G-morphism $f: X \to A_\chi$ for some model A_χ.*

Remark. The varieties A_χ are the only affine connected G-varieties which constist of two orbits one of which is a point. The proof of this proposition requires some new propositions.

Recall that we will denote by NS_G the variety of unstable points in G-module $\mathbb{C}^N$. In view of the Theorem 1 $NS_G = GNS_T$ for the fixed maximal torus $T \subset G$, because the maximal tori are all conjugate. The space $\mathbb{C}^N$ as a T-space can be decomposed into a direct sum of so called "weight subspaces" $\mathbb{C}^N = \sum_{\text{char } T} \mathbb{C}_\chi$ where T acts on $\mathbb{C}_\chi$ by means of χ.

Definition. For any vector $v \in \mathbb{C}^N$ we denote by $\operatorname{Supp} v$ the finite set of $\chi \in \text{char } T$ such that the projections of v into $\mathbb{C}_\chi$ are non zero.

For any set X we put $\operatorname{Supp} X = \bigcup_{v \in X} \operatorname{Supp} v$.

By $\langle \mathrm{Supp}\ v \rangle$ we denote the convex envelope in the linear real space $\mathcal{X}(T) \underset{\mathbb{Z}}{\otimes} \mathbb{R}$ of the set Supp v. It is a polytope.

Now we describe the variety NS_G.

<u>Lemma 1. Any vector</u> $v \in NS_G$ <u>is equivalent under</u> G <u>to some vector</u> s <u>with</u> $s \in \langle \mathrm{Supp}\ s \rangle$.

<u>Proof</u>. This follows at once from Theorem 1 and from the fact that $GNS_T = NS_G$.

So for each $l \in \hat{\mathcal{X}}(T)$ we can take the half-space $(l,x) > 0$ in $\mathcal{X}(T) \underset{\mathbb{Z}}{\otimes} \mathbb{R}$ and define the subspace $V_l \subset \mathbb{C}^N$ as the vector subspace which contains all vectors v such that Supp v half-space $(l,x) > 0$. Then $NS_G = \bigcup_l GV_l$. The number of different space V_l is finite in view of the finiteness of Supp $\mathbb{C}^N$.

<u>Lemma 2. Suppose that there is a linear P-invariant subspace</u> $V \subset \mathbb{C}^N$ <u>where</u> P <u>is a parabolic subgroup of</u> G. <u>Then the space</u> GV <u>is closed in</u> $\mathbb{C}^N$.

<u>Proof</u>. In fact GV is the image of the bundle $G \underset{P}{\otimes} V$ with fibre V on the compact manifold G/P. The morphism $h : G \underset{P}{\otimes} V \longrightarrow G$ is proper and $h^{-1}(x)$ is compact for any x ; thus GV is closed.

Now we define some submanifold which slightly generalize GV_l. For that let us take a scalar product $\langle , \rangle$ on $\mathcal{X}(T) \underset{\mathbb{Z}}{\otimes} \mathbb{R}$ which is positively definite and invariant under the action W(G) on $\mathcal{X}(T) \underset{\mathbb{Z}}{\otimes} \mathbb{R}$. As usual if the Lie algebra of G is simple then it is uniquely defined up to multiplication by a scalar. For each $\alpha \in \mathcal{X}(T) \underset{\mathbb{Q}}{\otimes} \mathbb{R}$ define the subset $\Delta_\alpha \subset \mathrm{Supp}\ \mathbb{C}^N$ by $x \in \Delta_\alpha$ iff $(x-\alpha , \alpha) \geqslant 0$.

Define $\partial \Delta_\alpha$ as the set $(x-\alpha, \alpha) = 0$, $x \in \mathrm{Supp}\ \mathbb{C}^N$. Consider only those values α for which $\partial \Delta_\alpha$ is non empty. Now define V_{Δ_α} as the space of vectors v such that Supp $v \subset \Delta_\alpha$, and $V_{\partial \Delta_\alpha}$ as the space of the vectors v such that supp $v \subset \partial \Delta_\alpha$

<u>Lemma 3. Consider the roots</u> r <u>of the Lie algebra</u> $\mathfrak{g}$ <u>with respect to the torus</u> T. <u>The space</u> V_{Δ_α} <u>is invariant under the</u>

parabolic subgroup. P_α whose the Lie algebra is generated by the eigenvectors of roots r such that $(r,\alpha)\geqslant 0$. We denote them by the same letter.

The proof is a direct consequence of the formula for action of the generators of a Lie group on $\mathbb{C}^N$.

So by Lemma 2, GV_α is a closed space for all α.

Remark. There are only a finite number of different submanifold GV_α in $\mathbb{C}^N$.

The group P contains a non trivial unipotent radical $U(P)$; this group is generated by $\exp(t\,r_i)$, where r_i is a root of $\mathfrak{g}$ and $(\alpha,r_i)>0$. In the same way P contains a reductive subgroup H which is generated by roots r_i with $(\alpha,r_i)=0$ and the torus T.

The group P can be represented as an extension

$$0\longrightarrow U(P)\longrightarrow P\underset{(s)}{\longrightarrow} M\longrightarrow 0 \quad .$$

For any $m\in\mathbb{Z}$ such that $m\alpha\in\mathcal{X}(T)$, $m\alpha$ is a character of the reductive group H. Define $H_0=\ker\alpha$ and consider the rings S of the H_0-invariant regular functions on $V_{\partial\Delta_\alpha}$. $V_{\partial\Delta_\alpha}$ is an H-invariant subspace of $\mathbb{C}^N$. Thus S has a direct decomposition into homogeneous subspaces S^m. The group H acts on S^m by mean of $m\alpha(H)$. Let us remark that in the above decomposition we have $m>0$ because for any $v\in S^m(V_{\partial\Delta_\alpha})$, supp v lies in the set $m\,(\mathrm{Supp}\,V_{\partial\Delta_\alpha})$ and for any point in $x\in\mathrm{Supp}\,V_{\partial\Delta_\alpha}$ we have $(x,\alpha)=(\alpha,\alpha)>0$.

Now we can construct a G-morphism from the fibre bundle $G\underset{P_\alpha}{\otimes}V_{\Delta_\alpha}$ into A_{m_α} by mean of the functions $f_m\in S^{m_\alpha}$. In fact f_m gives us a P-morphism

$$\hat{f}_{m_\alpha}: V_{\Delta_\alpha}\longrightarrow V_{\partial\Delta_\alpha}\approx V_{\Delta_\alpha}/V_{\Delta_\alpha\setminus\partial\Delta_\alpha}\xrightarrow{f_{m_\alpha}}\mathbb{C}_{m_\alpha}$$

and, as m_α lies in a "Weyl chamber" of P_α, we get the morphism as a G-extension of f_{m_α}. We call it Ψ_α.

Lemma 4. There is a commutative diagramm of G-morphisms

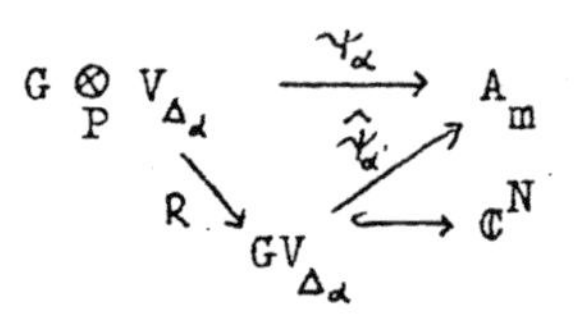

Proof. Let us take a Bruhat decomposition of any element $g \in G$, $g = pwu$, where $u \in U(P_\alpha)$, $w \in W(G)$, $p \in P_\alpha$. To prove the Lemma we need to check that if $v \in V_{\Delta_\alpha}$ and $gv \in V_{\Delta_\alpha}$ for some $g \notin P_\alpha$, then $\Psi_\alpha(gv) = g\Psi_\alpha(v)$ or, more precisely, that $\Psi_\alpha(v) = 0$, because the lines $\mathbb{C}_{m\alpha}$ and $g\,\mathbb{C}_{m\alpha}$ in $A_{m\alpha}$ cross only at the origin. We remark that $(pwu)v \in$ means that $w(uv) \in V_{\Delta_\alpha}$, but $\Psi_\alpha(uv) = \Psi_\alpha(v)$. So, if we put $uv = \hat{v}$, then we have to prove the folloäing: if $w\hat{v} \in V_{\Delta_\alpha}$ for any $w \notin W(P)$, the subgroup of the Weyl group leaving P invariant, then $\Psi_\alpha(\hat{v}) = 0$. Take a Veronese morphism $\nu_m : \mathbb{C}^N \to \mathbb{C}^{N \otimes m}$ and the induced morphism $\nu_m : V_{\partial\Delta_\alpha} \to S^m(V_{\partial\Delta_\alpha})$. The function f_α can be obtained as the composition of ν_m and the linear projection on one dimensional invariant subspace $\mathbb{C}_{m\alpha}$ of weight $\langle m\alpha \rangle$ in $S^m(V_{\partial\Delta_\alpha})$. We can consider the Weyl transformation as a linear transformation of the lattice $\mathcal{X}(T)$ and of the space $\mathcal{X}(T) \otimes_{\mathbb{Z}} \mathbb{R}$. For every $v \in \mathbb{C}^N$, $w(\mathrm{Supp}\ v) = \mathrm{Supp}\ (wv)$. Thus, if $v, vw \in V_{\Delta_\alpha}$, then $\mathrm{Supp}\ wv \subset \Delta_\alpha$. For any y we have $S^m y \in m\ \mathrm{Supp}\ y$. Therefore, if $f_{m\alpha}(v) \neq 0$, we have $\alpha \in \langle \mathrm{Supp}\ v \rangle$. As w is an isometry on $\mathcal{X}(T) \otimes_{\mathbb{Z}} \mathbb{R}$, we have $\langle \alpha, \alpha \rangle = \inf_{x \in \mathrm{Supp}\ v} \langle x, v$; but $\mathrm{Supp}\ w \in \Delta_\alpha$ and $\inf_{x \in \mathrm{Supp}\ v} (x, \alpha) \geq \langle \alpha, \alpha \rangle$, . So we have the equality $(\alpha, \alpha) = \langle w\alpha, \alpha \rangle$ and this implies that $w\alpha = \alpha$ and that $w \in W(P)$ in view of the fact that the set of the roots of P is w-invariant. Thus we get a contradiction ($f_\alpha(v) \neq 0$!). This proves that there is a morphism $\hat{\Psi}_\alpha$ such that $\Psi_\alpha = R\,\hat{\Psi}_\alpha$.

Remark. We can prove that for any f_α, for m large, the morphism induced by $f_\alpha^m = \hat{f}_{m\alpha}$ is regular morphism $GV_\alpha \to A_{m\alpha}$.

Now consider the subset of GV_α where all morphisms f_α are trivial.

Lemma 6. $\bigcap_{\substack{f \in S^m \\ m>0}} \hat{f}^{-1}_{m\alpha}(0)$ in GV_α is a finite union of sets $GV_\chi \cap GV_\alpha$ where $\chi \in \Delta_\alpha$, $(\chi,\chi) > (\alpha,\alpha)$.

Proof. The set where all $f_{m\alpha}$ are zero is exactly the manifold of H_0-unstable points in $V_{\cap \Delta_\alpha}$ but it can be described as $H_0 \hat{V}_s$, $s \in \chi(T \cap H_0)$ (see Lemma 1). So the full manifold where all f_α are zero in V_{Δ_α} is $H_0 V_s \cup V_{\Delta_\alpha \setminus \cap \Delta_\alpha}$ and the corresponding submanifold in $\mathbb{C}^N$ may be described as $G(V_{\alpha+\gamma}) \cap GV_\alpha$, $(\gamma,\alpha) > 0$.

This proves the existence of the morphism $f : X \to A_\chi$ when X is an irreducible G-manifold.

The case when $X = \bigcup_{i \geq 0} X_i$ may be reduced to the previous one by using the linear G-space $\mathcal{F}$ of the regular functions on X which are trivial on $\bigcup_{i>0} X_i$. The image $\mathcal{F}(X)$ and its closure $\overline{\mathcal{F}(X)}$ are irréducible G-manifolds. So, in view of the previous construction we have a morphism

$$f_{m\chi} : \overline{\mathcal{F}(X)} \to A_{m\chi}$$

which defines the morphism

$$\overline{f}_{m\chi} : X \to \overline{\mathcal{F}(X)} \to A_{m\chi} .$$

This proves our main theorem.

Example. Consider the case $G \approx SL(2,\mathbb{C})$ and an irreducible linear $SL(2,\mathbb{C})$-space $S^i\mathbb{C}^2$ of dimension $i+1$. $SL(2,\mathbb{C})$ acts on this space as on the space of the homogeneous polynomials of two variables x,y of degree $(i+1)$.

As it is well known (see D. Mumford: Geometric invariant theory) the unstable points correspond to the polynomials $p(x,y)$ which in some other coordinates x',y' can be written as $p(x',y') =$

$= x^{ik}\ p(x,y)$ where $k > [i/2]$. The line $\{x' = 0\}$ is defined uniquely by this property for any $p(x,y) \neq 0$. So our manifold $NS_{SL(2,\mathbb{C})}$ in $\mathbb{C}^{i+1}$ can be described as an image of a linear homogeneous bundle on $\mathbb{P}^1 = SL(2,\mathbb{C})/\Delta$ $\left(\Delta = \begin{pmatrix} a & b \\ 0 & a^{-1} \end{pmatrix} \right)$ where the fibre is the linear space $E^{(i)}$ of polynomials $p(x',xy)$ of degree less than $[i/2]$. The functions f_α can be choosen as the linear functions on $E^{(i)}$ corresponding to the coefficient of the monomial of highest degree in y.

3. Subbundles in cotangent bundle of a surface.

Now we apply the previous results to study the vector bundles on compact manifolds. Let us take a vector bundle E^k. We may assume that it is associated to a principal fibration X_γ with fibre $GL(k)$, $\gamma \in H^1(X, \mathcal{O}(GL(k)))$. Then for any representation $\rho : GL(k) \longrightarrow GL(N)$, we can construct a bundle $\tilde{E}^\rho$ on X associated to X_γ. Suppose that ρ is an irreducible representation and that the bundle $\tilde{E}^\rho$ has some non trivial section s. Now, taking any regular $GL(k)$-invariant function f on $\mathbb{C}^N \approx \tilde{E}^\rho_x$ we get a regular function on X, $f \circ s$. Because X is compact, $f \circ s$ must be constant and so, using our results of the theory of the morphisms into models, we get the following

<u>Proposition 2</u>. <u>If</u> s $H^0(X,\tilde{E}^\rho)$ <u>is stable at a point</u> x, i.e. <u>the point</u> s_x <u>is stable under the action of</u> $GL(k)$ <u>on</u> $\mathbb{C}^N$, <u>then</u> s <u>is stable at any other point</u> y. <u>Moreover there are two bases</u> $(e_1,\dots,e_k)$, $(l_1,\dots,l_k)$ <u>in</u> E_x <u>and</u> E_y <u>respectively and a polynomial</u> P <u>such that</u> $s_x = P(e_1,\dots,e_k)$ <u>and</u> $s_y = P(l_1,\dots,l_k)$.

<u>Corollary 1</u>. <u>If</u> $s_x = 0$ <u>for a point</u> $x \in X$ <u>then</u> s_y <u>is unstable for any</u> $y \in X$.

<u>Corollary 2</u>. <u>If</u> s_x <u>is stable and</u> $\dim GL(k)s_x$ <u>is equal to</u> $\dim GL(k)$ <u>then the bundle</u> E <u>taken on a finite unramified cove-</u>

ring $\widetilde{X}$ of X becomes a product $(\overset{k}{\underset{1}{\oplus}} \mathcal{O}) \otimes F$ where F is one dimensional.

These results can be obtained using the morphism into the first kind mostels, we get the following

Proposition 3. If s is semi-stable at $x \in X$ and $\overline{G s_x} \supset s \simeq G/H$ then s is everywhere semi-stable and the structure group of one bundle E reduces to H .

Now suppose that any triangular $GL(k)$-invariant function f wanishing at the origin also vanishes at s_x . Thus s_x is unstable at any point x X . No we can use the morphisms in the third kind models. We suppose only that X is normal and connected algebraic variety. Then s_x lies in a non linear subbundle $N\, S_{0,}^{GL(k)} \hookrightarrow \mathbb{C}^N$, subfibration of unstable $GL(k)$ points in E , and any non trivial morphism of this fibration into A_χ gives us a section of the bundle $A_{\chi,\gamma}$ on X . This bundle is the bundle of highest weight vectors in the linear bundle $E_{\chi,\gamma}$. Our previous assumptions on the irreducibility of ρ gives us that $\rho(Z)$, $Z \simeq \mathbb{C}^*$ being the central subgroup of $GL(k)$, is trivial and that the character χ is trivial on $\mathbb{C}^*$. So it can be expressed in the following way. Take a torus T , the maximal torus of $GL(k)$

$$T = \begin{pmatrix} \lambda_1 & & 0 \\ & \ddots & \\ 0 & & \lambda_k \end{pmatrix} , \qquad \chi = (n_1, \dots, n_k) .$$

We order them in increasing way so we have a natural flag of subspaces in E_x ; $E_x = E_0 \supset E_1 \ \dots \supset E_k$; $\operatorname{Supp} E_i \subset \operatorname{Supp} E_{i-1}$ and $\operatorname{Supp} E_{i-1} \setminus \operatorname{Supp} E_i$ is the set of the points with the same $\{n_i\}$ for the character. $\chi = (n_1, \dots, n_k)$ and n_i is minimal on $\operatorname{Supp} E_i$. This gives us a representation for a point in A_χ as

$$(\det E)^{n_0} \otimes (\det E_1)^{n_1} \otimes \dots \otimes (\det E_k)^{n_k} \cdot$$

or, in a more usual way, as the product of volume elements

$(\wedge^{k_i} E_i)^{n_i}$ of flag spaces. We remark that the parabolic subgroup leaving this flag invariant is just the group P_α of the construction of Ψ_α.

The tensor $\Psi_\alpha(s)$ defines a flag of subspaces $E_0 \supset E_1 \ldots \supset E_k$ at those points x where $\Psi_\alpha(s_x) \neq 0$ because it defines a parabolic subgroup of $GL(k)$. That parabolic subgroup is defined uniquely away from a subvariety of codimension greater than two. So on $X \setminus Y$ there is defined a flag of subbundles in E. This flag $E_0 \supset E_1 \ldots \supset E_k$ has an additional property. Our tensor $\Psi_\alpha(s)$ can be expressed as $\Psi_\alpha(s)$ $H^0(X, \bigotimes_1^k \det E_i^{r_i} \otimes \det E^{-l_i})$ where $r_i > 0$ and $l_i = \sum_1^k r_i$ rank E_i/rank E. In fact the point $\Psi_\alpha(s)$ fixes a line bundle $A_{\chi,\gamma}$ and a point on it. This means that we have a trivial line bundle $\mathcal{O}$ on X and a homomorphism $h: \mathcal{O} \to A_{\chi,\gamma}$. The image $h(\mathcal{O})$ lies in the one dimensional flag subbundle in A_χ defined by the flag $E \supset E_1 \ldots \supset E_k$.

As $\sum_i n_i = 0$ we get a representation

$$\det E_1^{r_1} \otimes \ldots \otimes \det E_k^{r_k} \otimes \det E^{-\sum_i l_i} \tag{1}$$

where $r_i = n_i - n_{i-1}$ 0.

This product equals to the product of formal bundles $\bigotimes_{i=1}^k$ (rank E $\det E_i \otimes$ (rank E_i $\det E)^{-1}$).

In fact, when we change our bundle E by a multiplication of one dimensional bundle F this product does not change. Let us take instead of F the formal bundle F such that $(F^*)^{\text{rank } E} =$ $= \det E$. Then (1) is equal to the product $\prod_{i=1}^k \det (E_i \otimes F)^{r_i}$, $\det (E_i \otimes F)^{r_i} = r_i$ rank E_i $(\det E_i \otimes F)$ and $F \approx \frac{1}{\text{rank } E} \det E^{-1}$.

<u>Remark</u>. This formal calculation with formal bundles, which does not exists in nature, help us to understand what happens in the general case.

We may also consider a bundle $\{E^{\otimes s} \otimes F\}$ where rank $F = 1$ as the tensor product of $(E \otimes F/\text{rank } E)^{\otimes s}$. So, even if the cor responding bundle does not exist, its tensor product exists. Now we give the definition of unstable bundle.

Definition. Let E be a vector bundle on a smooth variety X. We say that E is unstable if there exists a bundle E^{ρ} associated to the same principal fibration than E, such that: $\det E^{\rho} = 0$, $H^{0}(X, E^{\rho}) \neq 0$ and there is a section $s \in H^{0}(X, E^{\rho})$ which lies in a HV-subfibration of E^{ρ} and $s_x = 0$ at every point.

In view of the previous consideration it can be proved the following

Theorem 4. If there exists a bundle E^{ρ} such that $\det E^{\rho} = 0$, E^{ρ} is not trivial and $H^{0}(X, E^{\rho}) \neq 0$, but there is a section $s \in H^{0}(X, \widetilde{E}^{\rho})$ which is trivial in some point $x \in X$, then all not trivial bundles associates to E^{ρ} are unstable.

Remark. In the case of curves this definition coincides with that one given by Seshadri, in view of the fact that actually $\mathrm{Pic}\, X / \mathrm{Pic}^{0} X = \mathbb{Z}$ and therefore we may select a special flag containing one element only.

When $\mathrm{rank}\, E = 2$ the flag consist of one element, the one dimensional bundle F with a homomorphism $h : F \longrightarrow E$ which is a monomorphism away from a submanifold of codimension two. So that

1) h is a holomorphic morphism $F \longrightarrow E$
2) for some positive integer n, the divisor $n \cdot (2F - \det E)$ is effective.

Remark that if such subbundle F exists then it is uniquely determined. We shall use this property in all our considerations.

The proof of these facts becomes evident if we represent our bundle E, away from a finite number of points, as an extension

$$0 \longrightarrow F \xrightarrow{h} E \longrightarrow L \longrightarrow 0 .$$

Then $s^{2m} E \otimes (-m \det E)$ can be expressed as a sequence of extensions and L corresponds to a subbundle $F^{2m} \otimes (-m(F + E))$.

It follows that

$$f_{\chi}(s) \hookrightarrow A_{\chi,\gamma} \subset S^{2m} E \otimes (-m \det E)$$
$$f_{\chi}(s) \hookrightarrow L \approx m\,(2F \otimes (-(F+E))) \quad .$$

If there is another such a bundle $\tilde{F}$, then there is a morphism $F+\tilde{F} \longrightarrow \det E$ and $\dim H^0(X, 2\det E - 2(F+\tilde{F})) > 0$. As $\dim H^0(X, 2(F+\tilde{F}) - 2\det E) > 0$ we have $F+\tilde{F} \simeq E$. It follows that if $s_x = 0$ at every point then F is uniquely determined.

4. Symmetric tensors on a surface.

In this lecture I should like to discuss some ways to find tensors of special kind on a variety. We restricts ourselves to the case X is a smooth algebraic surface V .

Take a bundle E and its symmetric powers S^mE. We can calculate the Euler-Poincaré characteristic of this bundle: $\chi(V,S^mE) = \sum_{i=0}^{2} (-1)^i \dim H^i(V,S^mE)$. By the Riemann-Roch formula this is a polynomial of m of degree $k+1$, where $k = \operatorname{rank} E$ and the coefficients of this polynomial depends only on the Chern classes of the surface and the bundle E . Moreover this formula implies:

1) the coefficients of the highest power of m does not depend on invariant of the surface.
2) this coefficient is $c_1^2 - c_2$ multiplied by some constant $s(k)$ which depends only on the rank of E .

We give a brief survey of these calculations. Recall that

$$\chi(V,S^mE) \approx \operatorname{ch} S^mE \circ \operatorname{Td} V \; V$$

$$\operatorname{ch} S^mE = \operatorname{rank} S^mE + c_1(S^mE) + \frac{c_1^2(S^mE)}{2} - c_2(S^mE)$$

$$\operatorname{Td} V = (1 - \frac{k}{2} + \frac{k^2+\chi}{12}) \; .$$

Now we get the expression for $\chi_m = \chi(S^mE)$

$$\chi_m = \text{rank } S^mE \, \frac{k^2+\chi}{12} - \frac{c_1(S^mE)k}{2} + \frac{c_1^2(S^mE)}{2} - c_2(S^mE)$$

We remark that this expression naturally divides into three parts. The growth of the first when m tends to infinity is less or equal than $c\,m^{k-1}$. For the second $\{2\}$ we have inequality: $\{2\} \leq c\,m^k$. Only the third part can grow as $c\,m^{k+1}$ does not depend on the invariants of V, $c_1(S^mE)$ and $c_2(S^mE)$ may be expressed in terms of $c_1(E)$ and $c_2(E)$. The direct calculation (and it could be simply reduced to the case of bundles of rank two by formal arguments) give $\chi_m \sim s(k)\, m^{k+1}\, (c_1^2 - c_2)$.

<u>Corollary</u>. <u>If</u> $c_1^2(E) = 0$ <u>and if</u> $c_2(E)$ 0 <u>then</u> χ_m <u>grows as</u> $c\,m^{k+1}$.

We wish to deduce the existence of sections of S^mE from the growth of (S^mE).

<u>Lemma</u>. <u>If</u> χ_m <u>grows as</u> $c\,m^{k+1}$ <u>when</u> m <u>tends to infinity then either</u>

1) $\dim H^0(V, S^mE^k)$ <u>grows as</u> $a\,m^{k+1}$

<u>or</u>

2) $\dim H^0(V, S^mE^{k*})$ <u>grows as</u> $b\,m^{k+1}$.

<u>Proof</u>. $\chi_m \leq \dim H^0(V, S^mE) + \dim H^0(V, S^mE^* \otimes K)$. By Serre duality we get $H^2(V, S^mE)$ $H^0(V, S^mE^* \otimes K)$. But the dimensions of the groups $H^0(V, S^mE^* \otimes K)$ and $H^0(V, S^mE^*\quad)$ don't differ very much. For there is an effective divisor D with $D-K$ also effective and exact sequence

$$0 \to S^mE^* \otimes K{-}D \to S^mE^* \otimes K \to S^mE^* \otimes K_{|D} \to 0 .$$

Consider the corresponding cohomology sequence

$$0 \to H^0(V, S^mE^* \otimes K{-}D) \to H^0(V, S^mE^* \otimes K) - H^0(V, S^mE^* \otimes K_{|D}) \to \dots .$$

The growth of the right hand term is less than $c\,m^k$ as it is less then symmetric power of a bundle on a curve. So

$$\dim H^0(V, S^m E^* \otimes K-D) = \dim H^0(V, S^m E^*) + O(k) \quad .$$

Similarly $\quad h^0(V, S^m E^* \otimes \mathcal{O}(K-D)) = h^0(V, S^m E^*) + O(m^k)$.

<u>Corollary</u>. <u>If</u> $c_1^2 - \frac{2k}{k-1} c_2 > 0$ <u>then the bundle</u> E^k <u>on a surface</u> V <u>is unstable</u>.

<u>Proof</u>. In fact the inequality is equivalent to inequality $c_2 \frac{k-1}{2k} c_1^2 < 0$ which is equivalent to $c_2(E \otimes (-\frac{\det E}{k})) < 0$. This bundle $E_0 \approx E \otimes (-\frac{\det E}{k})$ does not exist but there exist its symmetric powers $S^{mk}(E_0) \approx \{S^{mk} E \otimes (-\det E^{\otimes m})\}$ and we can easily conclude from our inequality that $c\,m^{k+1} \leq \dim H^0(V, S^m E_0) + \dim H^0(V, S^m E_0^*)$. Now we can define a restriction of $H^0(V, S^m E_0)$ at any fibre E_x . Then, as $\dim S^m E_{0x} \sim \dim S^m E_x \leq c\,m^{k-1}$, we can conclude that there exist many sections of $H^0(V, S^m E_0)$ or $H^0(V, S^m E_0^*)$ vanishing at the point $x \in V$. All these sections s are unstable sections and they gives us unstability for bundle E^k on V .

<u>Corollary</u>. <u>If</u> rank $E = 2$ <u>and</u> $c_1^2 > 4\,c_2$ <u>then there exists a subbundle</u> F <u>of rank</u> 1 <u>such that</u> $\dim H^0(V, (2F - \det E)^{\otimes n}) > 0$ <u>for some</u> $n > 0$.

Now we shall have a deal with a Severi's Problem:

<u>Problem</u>. Does there exists a 1-connected surface with non zero group of holomorphic symmetric tensor $H^0(V, S^i \Omega^1)$ for some i ?

The main reason for investigation of this question was that the group $H^0(V, \Omega^1)$ is a homotopy invariant of surface because this group is isomorphic to a first cohomology group of V .

<u>Lemma</u>. <u>If</u> $K^2 > \chi$ <u>for a surface</u> V <u>of general type then</u> $\dim H^0(V, S^m \Omega^1) > c\,m^3 + A$, $c > 0$, A <u>some constant</u>.

<u>Proof</u>. In fact

$$\dim H^0(V,S^m\Omega^1) + \dim H^0(V,S^m\Omega^{1*}\otimes K) > c\,m^3 + A$$

but for $s\in(m-1)K$ we have inclusion $s: S^m\Omega^{1*}\otimes K \to S^m\Omega^1$ (simply tensoring by s). So $\dim H^0(V,S^m\Omega^1) > \dim H^0(V,S^m\Omega^{1*}\otimes K)$. This proves the lemma.

<u>Remark</u>. Later we shall prove that $H^0(V,S^{2m}\Omega^1\otimes\mathcal{O}(-mk)) = 0$ for a surface of general type.

Now we shall construct many examples of surfaces V with $H^0(V,S^i\Omega^1)\neq 0$ for some i 1 .

Take a 1-connected variety $M^{k+2}\subset \mathbb{P}^N$ of dimension $k+2$ and k hypersurfaces L_i which have transversal intersection with M^{k+2}. Then we have a surface $V\subset M^{k+2}$.

It is a connected surface in M^{k+2}. If the degrees of L_i are large enough and $\dim M^{k+2}\geqslant 4$ then a simply calculation gives us that $1<c_1^2/c_2<\frac{2k}{k+1}$. So all such surfaces have not trivial symmetric tensors.

<u>Remark</u>. We can get a more exact result. With this procedure we can obtain surfaces with $H^0(V,S^4\Omega^1)\neq 0$ if $\dim M^{k+2}\geqslant 26$.

Another direct way of getting surfaces with $H^0(V,S^i\Omega^1)\neq 0$ for any $i>1$ is the following. Let us take a three dimensional abelian variety A^3. Consider an involution σ on A^3 $(x\mapsto -x)$ with finite number of fixed points, points of order 2 on A^3, v_i. Let us take a surface V^2 in A^3 so that

1) V^2 is a smooth hyperplane section of A^3,
2) V^2 contains all points v_i and at every point v_i has a local equation in a local plane coordinates

$$z_i = \int\omega_i\ ,\qquad \omega_i\in H^0(V^2,\Omega^1)$$

$$z_0 = a_0z_1^3 + b_2z_2^3 + 0(m^3)$$

It can easily be seen that on the desingularisation M^2 of V^2/G we have the following:

1) there exists a tensor $dz_o^2 = \omega_o^2 \in H^o(M^2, S^2\Omega^1)$,
2) M^2 is 1-connected,
3) all tensors which exist on M^2 can be expressed in a form $\sum a_{ijk}\omega_o^{k_1} \otimes \omega_1^{k_2} \otimes \omega_2^{k_3}$ where $k_1 > k_2 + k_3$, $\sum_i k_i = 21$, a_{ijk} are constants.

Now we can even construct a family of surfaces V_t^2 such that

1) the group $H^o(V_o, S^i\Omega^1)$ is not trivial for any $i = 2\lambda - 1$,
2) the group $H^o(V_t, S^i\Omega^1) = 0$ for any i and $t \neq 0$.

We use the previous construction for V_o and for V_t by changing only the local representation at points v_i to $(\alpha z_o + \beta z_1 + \gamma z_2) = \sum \beta_i z_i^3 + O(m^3)$. If α, β, γ are general then $H^o(V, S^i\Omega^1) = 0$.

Remark. The analog of this procedure give us a tensor $s \in H^o(V, S^3\Omega^1)$ on a 1-connected surface.

5. Finiteness Theorems for surfaces of general type.

In this lecture we shall prove the following

Theorem 1. Let V be a surface and Ω^1 its cotangent bundle. Then for any line-bundle E with a not trivial morphism $h: E \to \Omega^1$ we have $\dim H^o(V, nE) < cn + \beta$.

Proof. The proof consist of several steps.

Step 1. Let $s \in H^o(V, nE)$ be a section. We will associate a section $\tilde{s} \in H^o(\tilde{V}, p^*E)$ to this section s where $p: \tilde{V} \to V$ is some finite ramified covering of V. Consider the subvariety $s(V) \subset nE$ and let $\varphi_n: E \to nE$ be the natural morphism of line bundle (which is given locally by $z \mapsto z^n$, z being the coordina te on a fibre). Let $\tilde{V}_1 = \varphi_n^{-1}(s(V)) \subset E$. The morphism φ_n restricted to V_1 defines a finite covering $\Psi : V_1 \to V$ ramified along the zero locus of the section s. Let $r : \tilde{V}_1 \to V_1$ be the inverse image of E on V_1 corresponding to the projection $\tilde{p} : \tilde{V}_1 \xrightarrow{r} V_1 \xrightarrow{\varphi_n} V$. We have a non trivial morphism $h^* : \tilde{E} \to \Omega^1_{\tilde{V}}$ and a section $h^*(\tilde{s}) \in H^0(\tilde{V}_1, \Omega^1_{\tilde{V}_1})$ where $\tilde{s} \in H^0(\tilde{V}_1, \tilde{E})$ is the natural section of $\tilde{E}$ corresponding to s. We remark that the tensor p^*s can be expressed as $\tilde{s}^n$ locally at a point $x \in \tilde{V}_1$ where the projection $\tilde{p} : \tilde{V}_1 \to V$ is non ramified.

Step 2. The line bundle E defines a one dimensional foliation E_s on V with a finite number of singular points. Any section of E gives us a holomorphic form on surface V. If s_1, s_2 are two holomorphic section and $\omega = s_1/s_2$, as we have $d\omega = 0$ the fonction ω is constant along the fibres of E_s. Now take any curve X which is transversal to E_s and let $i : X \to V$ the natural ambedding.

Step 3. The restriction morphism $H^0(V_1, nE) \xrightarrow{i^*} H^0(X, nE)$ is a monomorphism. In fact if $s_{1|X} = 0$ then for any section s_2 which is not equal to zero on X we have s_1/s_2 is constant along fibres of E_s ; but $s_1/s_2 = 0$ on X, therefore $s_1/s_2 = 0$ on V. So we get that i^* is a monomorphism. But for any line bundle F on a curve X we have $\dim H^0(X, nF) < cn + \beta$ and this proves the theorem.

Now we shall give a generalization of the theorem for varieties of any dimension.

Theorem 2. Let M^k be a projective variety and let Ω^i be the i-exterior power of its cotangent bundle where $k \geq i$. Then for any line bundle E with non-trivial homomorphisme $h : E \to \Omega^i$ we have $\dim H^0(M, nE) < cn^i + \beta$.

Proof. The proof includes the same steps as the previous one and step Q.

Step 0. We reduce our situation to the case when $\dim M = i+1$. For this we have to consider a pencil of varieties M^{i+1} through the one general enough Q of $\dim Q = i$. This variety has to be chosen transversally to the foliation induced on general variety M^{i+1} by a bundle F_s.

Step 1. We remark that if $F \hookrightarrow \Omega^i(M)$, $\dim M = i+1$, then $\Omega^i(M) \otimes (-K) = T(M)$. Thus F defines a foliation of dimension one on M. For any $s \in H^0(M^k, nE) \subset H^0(M^k, S^n\Omega^i)$ we construct in the same manner a ramified covering variety $\tilde{M}_1$ where p^*s has a root of degree n, $p^*s = \tilde{s}^n$, $\tilde{s} \in H^0(\tilde{M}_1, p^*E)$.

Step 2. Again s_1/s_2 for $s_i \in H^0(M^k, nE)$, is a rational function which is constant along the fibers of one dimensional foliation since $d\tilde{s}_i = 0$.

Step 3. As we supposed in the step 0, the submanifold Q is transversal to the foliation F_s. Now we have a monomorphism $H^0(M,nF) \xrightarrow{i^*} H^0(Q,nF)$ and by similar arguments, $\dim H^0(M,nF) < \dim H^0(Q,nF) < c\, n^i + \beta$.

Remark. Y. Miyaoha used the procedure of "extracting a root" for a general symmetric tensor s. It can be described as producing a family $\{D_i\}$ of n line bundles on ramified covering $\tilde{V}_1$ such that the tensor p^*s on $\tilde{V}_1$ $(p : \tilde{V}_1 \to V)$ can be described as $p^*s \in H^0(\sum_{i=1}^{n} D_i) \subset S^n\Omega^1$ with the help of symmetric product of homomorphisms $h_i : D_i \to \Omega^1$. Using this procedure and some calculations for symmetric tensors with coefficients in one dimensional bundles and the Bott-Baum inequality for subbundle $D_iK - D_i^2 \leq c_2$ he proves the Bombieri's inequality for surfaces of general type: $c_1^2 \leq 3\, c_2$.

Now we consider the case of varieties of dimension more then two.

Theorem 3. Let M be a variety of dimension n with $P_{ic}M^n = \mathbb{Z}$ whose canonical class is non negative. Then if h is a polarization of M, we have an inequality of Lascoux-Berger

$$(c_1^2 - \frac{2n}{n-1} c_2) \wedge h^{n-2} \leq 0 .$$

In fact we can construct a surface $V \subset M$ with the following properties:

(i) Pic $V = \mathbb{Z}$ and is generated by the restriction of the generator of Pic M.

(ii) for any bundle $\{nh\}$, $n > 0$, we have the equality
$H^0(V, (-nh) \otimes \Omega^i|_V) = H^0(M, (-nh) \otimes \Omega^i)$.

(iii) $H^0(V, \Omega^i|_V) = H^0(M, \Omega^i)$.

From the instability criterion we obtain the instability of the bundle $\Omega^1|_V$ if the inequality $c_1^2 - \frac{2n}{n-1} c_2 > 0$ holds for a surface V.

I recall that there exists a flag of subbundles $F_0 \overset{h_0}{\hookrightarrow} F_1 \overset{h_1}{\hookrightarrow} \dots \hookrightarrow F_k \longrightarrow E$ where $h_i : F_i \to F_{i+1}$ is a monomorphism out of a finite number of points and for some integer vector $(n_0, \dots, n_k)$ the divisor $\bigoplus_i n_i (\det F_i - \frac{\operatorname{rank} F_i}{n} K)$ is effective. As the Picard group of M is $\mathbb{Z}$ we see that for some i, $n_i (\det F_i - \frac{\operatorname{rank} F_i}{n})$ is effective with $n_i > 0$. But the bundle $\det F_i$ has a non trivial morphism into Ω^i. Further $\det F_i$ is ample and from the previous theorem there is an inequality $\dim H^0(M, nF) \leq c n^i + \beta$. The condition (ii) talls us that a homomorphism $\det F_i \to \Omega^i|_V$ along V extends to M. So we obtain a contradiction and get required inequality for M.

Remark. Our result is slightly weaker than the Lascoux-Berger conjecture for the case Pic $M = \mathbb{Z}$. Indeed we have obtained only the inequality $(c_1^2 - \frac{2n}{n-1} c_2) \wedge h^{n-2} \geqslant 0$ but not > 0.

The same argument can be used in the case Pic $M \approx \mathbb{Z}$ and $-K$ is ample if we assume that $-K$ is a generator of Pic M. In fact taking the surface $V \subset M$ which satisfies the properties (i), (ii) (iii) we get that the divisor n_i (det $F_i - \frac{\text{rank } F_i}{n} K$) is effective. This implies det $F_i \geqslant 0$ in Pic M. From (ii) and (iii) we get that $H^0(M, \Omega^i) \neq 0$, but for coniugation isomorphism we have $H^0(M, \Omega^i) \approx H^i(M, \mathcal{O}) = 0$ for $i > 0$ and this proves the result.

6. Foliations on surfaces.

In this lecture we will study the following problem. Let V be a surface of general type and consider the set of all curves of geometric genus g which can be birationally embedded into V. These curves lie in a number of connected algebraic families. We want to describe these families under some homological restrictions. We shall deal with a surface V, $K^2 > 0$ and rank Pic $V \geqslant 2$. Consider the group $H^0(V, S^i \Omega^1 \otimes F)$ where F is some linear bundle on V. As usual we consider the formal equality $S^i \Omega^1 \otimes F = S^i(\Omega^1 \otimes F/i)$. From a different point of view the above group is isomorphic to the group $H^0(P(T), i(D + F/i))$ where $P(T)$ is the projectivization of the tangent bundle $T(V)$ and $-D$ is a divisor called the Grothendieck generator of the group of divisors Pic $P(T)$ over Pic V. It corresponds to the tautological line bundle on $P(T)$. For any smooth curve X which has a non trivial morphism f into V we have the corresponding tangential morphism $t_f : X \to P(T)$ which maps a point x X to the pair $(f(x), (df)_x)$ in $P(T)$. Because of $\dim X = 1$, t_f is a holomorphic morphism into $P(T)$. Consider now a line bundle F with $F^2 < 0$ and $F \cdot K \geqslant 0$, where we take a product in the group Pic $V \underset{\mathbb{Z}}{\otimes} \mathbb{R}$ and suppose that for some rational q the expression $c_1^2 - c_2$ for a formal bundle $\Omega^1 \otimes qF$ is positive. Using the formula for Chern classes, we get $K^2 - e + 3K \cdot F\, q + 3F^2 q^2 > 0$ where $-e$ is the Euler characteristic of V. Then we have an asymptotic inequality

$$\dim H^o(V,S^m(\Omega^1 \otimes qF)) + H^o(V,S^m(T \otimes (-qF))) > cm^3 .$$

Suppose also that $(K+2qF)^2 > 0$ and consider the group $H^o(V,S^m(T \otimes (-qF)))$.

Lemma. The group $H^o(V,S^m(T \otimes (-qF)))$ is trivial.

Proof. In fact $H^o(V, n(-K-qF)) = 0$. Thus any tensor $s \in H^o(V,S^m(T \otimes (-qF)))$ is unstable on V .

So there exists a subbundle L of rank one with non trivial homomorphism into Ω^1 such that $\dim H^o(V,n_1 L \otimes (-K-2qF)n_2 \otimes -n_3 K) > 0$ for some positive (n_1,n_2,n_3). Then $H^o(V,nL) > c\,n^2$ and we obtain a contradiction.

So $\dim H^o(V,S^m(\Omega^1 \otimes qF))$ grows as $c\,m^3$. By the Iitaka theorem we get, for some m large enough that rational morphism $\nu_m : P(T) \to \mathbb{P}^N$ associated to the linear system $H^o(V,m(D+qF))$ is a birational embedding. Suppose that $B = B_F$ is the base divisor of ν_m .

Definizion. A curve X is said regular if $t_f(X) \subset B$.

Let X be a regular curve. Then the degree of $\nu_m\, t_f(X)$ is less than the degree of m-symmetric tensors on X if $X \cdot F \leq 0$. In fact $\{D \cap F\} \leq mg$ and $F \cdot X \quad 0$. So all regular curves of geometric genus g are mapped under a morphism ν_m to a set of curves of bounded degree in $\mathbb{P}^N$. Thus they lie in an algebraic family. We now consider the values of the homological class $f\,X$ for which we can find such bundle F . We have to consider only the case when Pic V is two dimensional because both inequalities $X \cdot F \leq 0$ and $K^2 - e + 3K \cdot F/m + 3F^2 \frac{1}{m^2} > 0$ are invariant under the compact group of transformation $SO(m)$ which leaves the form $\langle x,x \rangle$ on Pic $V \otimes_{\mathbb{Z}} \mathbb{R}$ invariant and vector bundle K fixed. The corresponding quadratic form on the 2-plane (x,y) has the form x^2-y^2 . If $K^2 - e > 0$, then the trivial bundle $F = 0$

satisfies our criterion and so we have the following

Lemma. If $K^2 - e < 0$ then the family of regular curves for $F = 0$ of geometric genus g is algebraic.

Now suppose that $K^2 - e < 0$. We attempt to find the set in Pic V for which there exists such a bundle, perharps different for different curves. Geometrically we have an equation of a hyperbola in plane:

$$(1) \qquad K^2 + 3q\, K \cdot F + 3q^2 F^2 - e = 0 \quad \text{or} \quad \frac{3}{4}(K+2qF)^2 = e - \frac{1}{4} K^2 .$$

The hyperbola has as an asymptotic lines two lines parallel to the cone $\{X^2 \cdot 0\}$, L_1, L_2, and the cone K_1 of bundle F with the inequality (1) is tangent cone at the origin. So if we take F_1 in K_1 we can find a rational number q such that $K^2 + 3\, K \cdot F\, q + 3F^2 q^2 - \chi > 0$ and for X which has $F \cdot X \leq 0$. Thus we have the following

Lemma. The family of all regular curves of genus g which don't lie in a half plane $F \cdot X > 0$ is algebraic.

Consider the family $\mathcal{F}$ of all regular curves of genus g whose homological class $f(X) \in H_2(M, \mathbb{Z})$ is not contained in a cone $K^{\varepsilon}_{D_1}$ which is dual to a smaller cone $K^{\varepsilon}_1 = (3/4 - \varepsilon)(K^2 + 2qF)^2 > \chi - \frac{1}{4} K^2$.

Then $\mathcal{F}$ is an algebraic family since it is the union of algebraic families for a finite number of the bundles F_i .

So we have proved the following theorem:

Theorem. The family of curves of genus g constisting of all curves X with $f(X) \notin K_{D_1}$ is an algebraic family.

Corollary. There are only a finite number of homological classes $[f(X)]$ where geometric genus of X is g and X is regular for any F in the set $\{ \text{Pic } V \otimes_{\mathbb{Z}} R \setminus K_{L_1}$.

This gives us the Mordell problem for the surfaces of general type. Now we describe the irregular curves on V. Suppose that $t_f(X) \subset B_F$ a base locus divisor on $P(T)$ for $i(D+qF)$. Now B_F defines a finite number of direction at any point $x \in V$ which does not lie in a linear component of divisor B_F. Thus there exists a finite number of surfaces $B_i \subset B_F$ which map surjectively to V via $p: P(T) \to V$. Any B_i corresponds a linear subbundle l_i of the tangent bundle of the desingularization surface $\tilde{B}_i$ in such a way that if $t_f(X) \subset B_i$, then $t_f(X)$ is an integral curve on $\tilde{B}_i$ of the bundle l_i. So we obtain the following proposition

Proposition. All irregular F-curves are the images of curves which come from integral curve of finite number of foliations on finite of finite number of foliations on finite number of surfaces $\tilde{B}_i$.

So we have to describe the integrable curves of foliation if we want to describe all curves of genus g.

7. The VII$_0$ surfaces with $b_2 = 0$.

In this lecture we shall describe F-irregular curves. It follows from the previous lecture that all such curves lie in a finite number of foliations on ramified covering V of our surface V. So we are interested in integral curves of a foliation. We recall that a one-dimensional foliation on a surface is the same as a rank one subbundle of Ω^1 or of the tangent bundle of V. Suppose that two curves X_1, X_2 are integral for a foliation which is defined by $h: F \to \Omega^1$, where rank h is less than 1 only at a finite number of points p_i. Then the only intersection points of these two curves X_i can be the points p_i. The main theorem which we shall prove in the following

Theorem. Let us take a foliation F_S on a surface V. Then it must satisfy either

1) <u>there exists a rational morphism</u> $p : V \to X$ <u>onto a curve</u> X <u>such that</u> F_S <u>is tangent to the generic fibre</u> of p.
2) <u>there exists only a finite number of algebraic curves which are integral for</u> F_S.

The proof is based on the local description of the foliation. Locally we may suppose that the foliation can be given by a holomorphic form $\omega = P\,dx + Q\,dy$ with only one singular point namely the origin. Let us take local algebraic curve in a small neighbourhood U of the origin which are integral for ω. Thus we have $\omega/X = 0$.

Remark. This definition is somewhat different from the normal definition of integral curve of a foliation but they differ only on a divisor.

We shall call such a locally irreducible curve a <u>local algebraic integral</u> and denote it [l.a.i.]. Now we use the procedure of the monoidal transformation. Take for U a surface $\tilde{U}$ which is obtained by a number of σ-processes above the origin. Then if $p . \tilde{U} \to U$ is the corresponding morphism, we can define the induced foliation on $\tilde{U}$ by means of the form $p^*\omega$. Again it has only finite number of singular points. The [l.a.i.] of the induced form $p^*\omega$ are the same as that of ω except for the exceptional components. We have the following

<u>Theorem</u>. <u>There exists a surface</u> $\tilde{U}$ <u>over</u> U <u>such that any two</u> [l.a.i.] $p^*\omega$ <u>on</u> $\tilde{U}$ <u>which don't lie in</u> $p^{-1}(0)$ $(p : \tilde{U} \to U)$ <u>are disjoint</u>.

The proof of this is rather technical and we don't go through.

We apply the theorem to the compact surface V and foliation F. Making a finite number of σ-processes we get a surface $\tilde{V}$ and a foliation F_S on it such that the non-exceptional integral algebraic curve not have no intersection. Thus for any two such curves we have $X_i \cdot X_j = 0$.

Let us take integral curves with $X_i^2 = 0$. The number of them is less than rank $\operatorname{Pic}\tilde{V}$ and it is finite. All the curves with $X_i^2 = 0$ are homologically linearly equivalent because on the plane $(X_1, y) = 0$ the form X^2 is negatively definite and if $y^2 = 0$ then $y = tX$. So if $\operatorname{Pic}^\circ V = 0$ then we get a pencil $n_1X_1 \sim n_2X_2$ containing all our curves. If $\operatorname{Pic}^\circ V \neq 0$ then we get a pencil with the base curve of genus more than 0. So either our foliation is tangent to the pencil of curves or our integral curves lie in a divisor of this pencil and therefore coincide with them. This proves the theorem.

CENTRO INTERNAZIONALE MATEMATICO ESTIVO

(C.I.M.E.)

PLURICANONICAL MAPPINGS OF SURFACES WITH $K^2 = 1,2$, $q=p_g=0$

F. CATANESE

PLURICANONICAL MAPPINGS OF SURFACES WITH $K^2 = 1,2$, $q=p_g=0$ (*)

F. Catanese
Pisa (**)

I. Introduction.

This lecture is a continuation of Dolgacev's ones on surfaces with $q=p_g=0$, and considers those minimal models of such surfaces for which $K^2=2$ (numerical Campedelli surfaces) and those for which $K^2=1$ (numerical Godeaux surfaces): they are of general type by classification of surfaces.

The Main Theorem of [1] (to which we will refer as [CM]) asserted among other things that for a minimal surface of general type, Φ_{mK} denoting the rational map associated to the complete linear system $|mK_S|$, Φ_{mK} is birational, for $m \geq 3$, with these exceptions

---

(*) This seminar is an exposition of joint work of E. Bombieri-F. Catanese.

(**) This author is a member of G.N.S.A.G.A. of C.N.R..

a) $K^2=1$, $p_g=2$, $m=4,3$

b) $K^2=2$, $p_g=3$, $m=3$ and, possibly,

c) $K^2=1$. $p_g=0$, $m=4,3$, $K^2=2$, $p_g=0$, $m=3$.

It was later shown that the exceptions of c) don't really occur: the case $K^2=1$, $p_g=0$ $m=4$ was proven by Bombieri (unpublished) and subsequently by us along a simpler line of proof ([4]), the case $K^2=1$, $p_g=0$, $m=3$ by Miyaoka [6] and subsequently by Kulikov and us (along a different line of proof, unpublished), the case $K^2=2, p_g=0$, $m=3$, by Peters ([10]) in the particular case of a Campedelli double plane, in the general one by us ([3]) and later, independently by X. Benveniste (unpublished).

The main goal of this lecture is by one side to prove these results in the simplest fashion and by the other one to exhibit the application of some new lemmas (of [3]) which allow one to handle reducible curves in nearly the same way than non singular ones. We will give our proof for the first case, for the second we will give the main steps (in which our differs from Miyaoka's proof): for numerical Campedelli surfaces, finally, we remark that the proof appearing here is a combination of our with an argument of Benveniste's proof.

II. Some auxiliary results.

Lemma 1. On a surface S of general type with $K^2 \leq 2$, $q=0$ there is only a finite number of irreducible curves C with $K \cdot C \leq 1$.

Proof. Observe that if $C^2 < 0$ C is isolated in its class of numerical equivalence, hence in this case it suffices to show that the number of such classes is finite. Here we use the index theorem ([9] page 128) to the effect that on the subspace of numerical classes orthogonal to K the intersection form is negative definite: if $K \cdot C=0$ $C^2=-2$ (as $C^2<0$ and $KC+C^2=$ $=2p(C)-2 \geq -2$) and the number of such classes is finite (moreover such curves are numerically independent, see [5] pag. 177, [C.M.] pag. 174-5). If $K \cdot C=1$, then $(K-(K^2)C)$ is orthogonal to K, hence $0 \geq (K-(K^2)C)^2 = K^2((K^2)C^2-1)$ and so $C^2 \leq 1/K^2$; however $2p(C)-2 = 1+C^2 \geq -2$ implies C^2 odd, $C^2 \geq -3$, so $C^2<0$ unless only if it is $K^2=1$, $C^2=1$, K homologous to C.

Note that $(K-(K^2)C)$ belongs to a numerical class orthogonal to K with selfintersection bounded from below by -14, hence can belong only to a finite number of classes, and the same then occurs for C; finally if C is homologous to K $h^0(\mathcal{O}_S(C))=1$ (compare [6] or Dolgacev's lecture), and, the surface being regular, there is a finite number of such curves.

Q.E.D.

We refer to [3] for the proof of the following lemmas A, B, B'.

Lemma A. Let C be a positive divisor on a smooth surface S, $\mathcal{L}$ an invertible sheaf on C with $h^0(C,\mathcal{L}) \geq 1$: then either

i) there exists a section s not vanishing identically on any component of C, and $\deg_C \mathcal{L} \geq 0$, equality holding iff $\mathcal{L} = \mathcal{O}_C$

or

ii) there exists a section σ, $C_1, C_2 > 0$ such that $C=C_1+C_2$, $\sigma|_{C_1} \equiv 0$ but $\sigma|_{C'} \not\equiv 0$ if $C_1 < C' \leq C$, and $C_1 \cdot C_2 \leq \deg_{C_2}(\mathcal{L} \otimes \mathcal{O}_{C_2})$.

Lemma B. If Γ is an irreducible Gorenstein curve and $|\omega_\Gamma| \neq \emptyset$, then $|\omega_\Gamma|$ has no base points.

More generally a reduced point p of a curve C on a smoth surface is not a base point of $|\omega_C|$ if either

i) p is simple on C and belongs to a component Γ with $p(\Gamma) \geq 1$

or

ii) p is singular and for every decomposition $C=C_1+C_2$ $(C_i > 0)$ one has $C_1 \cdot C_2 > (C_1 \cdot C_2)_p =$ intersection multiplicity of C_1, C_2 at p.

Remark. If C is given by two elliptic curves meeting transversally at a point p, p is a base point of $|\omega_C|$, and in fact condition ii) is violated; however if C is given by more than three lines in the projective plane all meeting in the same point p, condition ii) is violated but p is not a base point. On a numerical Godeaux surface if ε is a torsion class $\neq -\varepsilon$, (D_ε denoting the unique curve of $|K+\varepsilon|$), $C=D_\varepsilon + D_{-\varepsilon}$ has $b=D_\varepsilon \cap D_{-\varepsilon}$ as a base point of $|\omega_C|$, and infact $D_\varepsilon \cdot D_{-\varepsilon} = 1 =$ $= (D_\varepsilon \cdot D_{-\varepsilon})_b$ (compare Dolgacev's lecture and the following of this).

Lemma B'. If p is a reduced singular point of a curve C lying on a smooth surface, denote by $\mathfrak{M}_p$ the maximal ideal of p in C, and let $\pi:\tilde{C} \to C$ be a normalization of C at p. Then $\mathrm{Hom}(\mathfrak{M}_p, \mathcal{O}_{C,p})$ can be embedded in the ring A of regular functions of $\tilde{C}$ at $\pi^{-1}(p)$.

Lemma 2 (X. Benveniste). Let S be a numerical Campedelli surface and m a positive integer: then the family $\mathfrak{F}_m$ of irreducible curves E such that $K\cdot E \leq m$ and $|E| = \{E\}$ is a finite one.

Proof. If $E \in \mathfrak{F}_m$, $h^2(E)=h^\circ(K-E) \leq h^\circ(K)=0$, hence R.R. gives $1+\frac{1}{2}(E^2-KE) = \chi(\mathcal{O}(E)) \leq 1$, so $(\#)$ $KE \geq E^2$. Associate to $E \in \mathfrak{F}_m$ the following numerical class ξ_E orthogonal to K: $\xi_E = 2E - (K\cdot E)K$.

Here, as in lemma 1, we use the index theorem to infer $\xi_E^2 \leq 0$. But $\xi_E^2 = 2E\cdot\xi_E = 4E^2 - 2(K\cdot E)^2$, and this, together with the already used inequality $E^2 \geq -KE-2$, gives the result that $\xi_E^2 \geq -2(K\cdot E)^2 - 8 - 4K\cdot E \geq -(8+4m+2m^2)$; in turn this implies that ξ_E may belong only to a finite set of numerical classes.

Suppose now $\xi_{E_1} \sim \xi_{E_2}$: if we prove that then either $E_1 \sim E_2$ or $E_1 \sim K+E_2$ we are done (the surface being regular each class can be given by at most 2m such curves where m is the order of the torsion subgroup T of $\mathrm{Pic}(S)$).

$\xi_{E_1} \sim \xi_{E_2}$ can be read as $2(E_1-E_2) \sim (KE_1-KE_2)K$ and we can assume $r = KE_1 - KE_2 > 0$; more over r is an even number, because $r(K\cdot E_i) = 2(E_1-E_2)\cdot E_i$ (r odd would imply $K\cdot E_1, K\cdot E_2$ to be even, which is obviously absurd).

This equality, in turn, when $i=1$, can be read as $2E_1^2 = 2E_1E_2 + r\,K\cdot E_1$ and the fact that $E_1\cdot E_2 \geq 0$

$$(\#)\quad E_1^2 \leq KE_1 \text{ soon imply } \begin{cases} E_1 \cdot E_2 = 0 \\ r = 2 \\ E_1^2 = K \cdot E_1 \end{cases}, \text{ finally } 2(E_1 - E_2) \sim 2K \text{ i.e. } E_1 \sim E_2 + K .$$

Q.E.D.

III. Birationality of Φ_{4K} for a numerical Godeaux surface.

Let S be a numerical Godeaux surface (a minimal model, as we always assume). We recall that for any S of general type, and $m \geq 2$ $P_m = h^0(\mathcal{O}(mK)) = \frac{1}{2} m(m-1)K^2 + \chi(\mathcal{O})$ ([5] pag. 184, [C.M.] pag. 185), so that in this case $P_2=2$, $P_3=4$, $P_4=7$.

Take U to be the Zariski open set whose complement is given by the union of the curves D such that $K \cdot D \leq 1$, with the locus of base points of $|2K|$ and of singular points of curves $C \in |2K|$.

Remark indeed that Φ_{4K} restricted to U is a regular map, then we claim

Theorem 1. $\Phi_{4K}|_U$ is an injective morphism.

Proof. Suppose that x,y are two points of U such that $\Phi_{4K}(x) = \Phi_{4K}(y)$; by our choice of U we may choose a curve $D \in |2K|$ s.t. $y \notin D$, and the unique curve $C \in |2K|$ s.t. $x \in C$: now $C+D$ is a curve of $|4K|$ passing through x, hence $y \in C$, and x,y are simple points of C.

Given a sheaf $\mathcal{F}$, denote by $\mathcal{F}(-x) = \mathcal{F}\mathfrak{m}_x$ (where $\mathfrak{m}_x$ is the ideal sheaf of the point x), and by $\mathbb{K}_x$ the sheaf supported at x with stalk $\mathbb{K}$. We obtain that $H^{\circ}(C, \mathcal{O}_C(K))=0$ by the exact sequence $0 \longrightarrow \mathcal{O}_S(-K) \longrightarrow \mathcal{O}_S(K) \longrightarrow \mathcal{O}_C(K) \longrightarrow 0$ together with the vanishing of $H^1(\mathcal{O}_S(-K))$ (this is by the vanishing theorem of Mumford, [8], asserting that if $\mathcal{L}$ is an invertible sheaf such that for large n $\mathcal{L}^n$ is spanned by global sections and has three algebraically independent sections, then $H^1(\mathcal{L}^{-1})=0$: it can be applied to $\mathcal{O}_S(mK)$, $m \geq 1$, and for later use we observe that by duality also gives $h^i(\mathcal{O}_S(mK))=0$ for $i \geq 1$, $m \geq 2$).

We will derive a contradiction by showing that $\mathcal{O}_C(K)$ is isomorphic to $\mathcal{O}_C(x+y)$.

Consider for this the following exact sequences:

$$0 \longrightarrow \mathcal{O}_S(4K-x-y) \longrightarrow \mathcal{O}_S(4K) \longrightarrow \mathbb{K}_x \oplus \mathbb{K}_y \longrightarrow 0$$

$$0 \longrightarrow \mathcal{O}_S(2K) \longrightarrow \mathcal{O}_S(4K-x-y) \longrightarrow \mathcal{O}_C(4K-x-y) \longrightarrow 0 .$$

Then from their cohomology sequences one gets $1=h^1(\mathcal{O}_S(4K-x-y))=h^1(\mathcal{O}_C(4K-x-y))$, and by Serre duality on the curve C $h^{\circ}(C,\mathcal{L})=1$, where $\mathcal{L}=\omega_C \otimes \mathcal{O}_C(4K-x-y)^{-1}=\mathcal{O}_C(x+y-K)$.

We are ready to apply lemma A, after observing that $\deg_C \mathcal{L} = 2-K\cdot C=0$; then if i) occurs $\mathcal{L} \cong \mathcal{O}_C$, what we wanted to show. Case ii) cannot occur: infact one would have $C=C_1+C_2$, $C_1 \cdot C_2 \leq \deg \mathcal{O}_{C_2} \otimes \mathcal{L} = -KC_2+$ (number of points of $C_2 \cap \{x,y\}$).

But by our choice of U, if x, or $y \in C_2$, then $KC_2 \geq 2$, so in any case $C_1 \cdot C_2$ should be non positive, and this contradicts the following result of Bombieri about connectedness of divisors homologous to pluricanonical divisors ([C.M.] pag. 181) : if $D \sim mK$, $m \geq 1$ and $D=D_1+D_2$, $D_i > 0$, then $D_1 \cdot D_2 \geq 2$, except when $K^2=1$, $m=2$ but then $D_1 \sim D_2 \sim K$.

Q.E.D.

IV. Birationality of Φ_{3K} for a numerical Campedelli surface.

By the just quoted formula here $P_2=3$, $P_3=7$.

From now on we suppose that x,y are two points such that $\Phi_{3K}(x) = \Phi_{3K}(y)$: as $P_2=3$ there exists a curve $C \in |2K|$ containing them both.

Lemma 3. $h^1(\mathcal{O}_C(3K-x-y)) = 2$.

Proof. The cohomology sequences of

$$0 \longrightarrow \mathcal{O}_S(3K-x-y) \longrightarrow \mathcal{O}_S(3K) \longrightarrow \mathbb{K}_x \oplus \mathbb{K}_y \longrightarrow 0$$

$$0 \longrightarrow \mathcal{O}_S(\underset{\substack{\text{III} \\ K}}{3K-C}) \longrightarrow \mathcal{O}_S(3K-x-y) \longrightarrow \mathcal{O}_C(3K-x-y) \longrightarrow 0$$

give $h^1(\mathcal{O}_C(3K-x-y)) = h^1(\mathcal{O}_S(3K-x-y)) + h^2(\mathcal{O}(K)) = 1+1$, as $h^1(\mathcal{O}_S(3K)) = 0$ and $\Phi_{3K}(x) = \Phi_{3K}(y)$.

Q.E.D.

Proposition 4. For general x,y , and $C \in |2K-x-y|$, x and y are simple points of the curve C which is hyperelliptic having $h_C = \mathcal{O}_C(x+y)$ as its hyperelliptic invertible sheaf.

Proof. The second part of the statement is an easy consequence of the first part, lemma 3 plus Serre duality on C , the first will be proven in two steps.

Step I: x belongs to only one irreducible component of C , the same holds for y .

In fact if, say, x belongs to two components Γ_1, Γ_2 of C , by lemma 1 $K \cdot \Gamma_i \geq 2$, hence is equal to 2, and by lemma 2

$h^{\circ}(\mathcal{O}_S(\Gamma_i)) \geq 2$. Write $C = \Gamma_1 + \Gamma_2 + F$ $(K \cdot F = 0)$ and consider that $|2K| \supset |\Gamma_1| + |\Gamma_2| + F$, $\dim |2K| = 2$: from this we deduce that $\Gamma_1 \equiv \Gamma_2$. In fact if Γ_1' is an irreducible curve $\in |\Gamma_1|$ there exist by the previous remark Γ_2', $\Gamma_2'' \in |\Gamma_2|$ such that $\Gamma_1 + \Gamma_2'' = \Gamma_1' + \Gamma_2'$, and so $\Gamma_2'' = \Gamma_1' + H_1'$, $\Gamma_2' = \Gamma_1 + H_1$, $H_1' \equiv H_1$, $H_1 \cdot K = 0$; but this implies $H_1' = H_1$ and so H_1 would be a fixed part of $|\Gamma_2|$: then H_1 must be 0 and $\Gamma_1 \equiv \Gamma_2$.

If it were $\Gamma_1 \not\equiv \Gamma_2$ x would be a base point of $|\Gamma_1|$, which cannot hold for general x (the numerical class of Γ_1 can range in a finite set); while if it were $\Gamma_1 = \Gamma_2$ one could take, by what has just been said, a curve $\Gamma_1' \in |\Gamma_1|$ not passing through x, and then consider $C' = \Gamma_1 + \Gamma_1' + F$.

Step II: x and y are simple points of C.

For this we can use lemma 3 and Grothendieck duality to infer that $\dim_{\mathbb{K}} \operatorname{Hom}(\mathfrak{m}_x \mathfrak{m}_y, \mathcal{O}_C) = 2$. Taking $\tilde{C}$ a normalization of C at x,y we observe that by step I $\tilde{C}$ is connected exactly as C, hence by Ramanujam's result ([11], lemma 3) $h^{\circ}(\mathcal{O}_{\tilde{C}}) = 1$. We can apply lemma B': x,y both singular would imply $h^{\circ}(\mathcal{O}_{\tilde{C}}) \geq 2$, a contradiction, while if however x is simple, y singular, one gets $h^{\circ}(\mathcal{O}_{\tilde{C}}(x)) = 2$ so x belongs to a rational curve; S being of general type, this cannot occur for general x.

Q.E.D.

From now on we suppose $C \in |2K-x-y|$ to satisfy the requirements of proposition 4.

<u>Lemma</u> 5. $\omega_C \cong h_C^{\otimes 6}$

<u>Proof</u>. If $s_1,\ldots s_6$ are six distinct general sections of $H^{\circ}(C,h_C)$ and $\operatorname{div}(s_i) = a_i+b_i$, one has that a section of $H^{\circ}(\omega_C)$ vanishing at $a_1,\ldots a_6$ vanishes at $b_1,\ldots b_6$ too.

Q.E.D.

We can pass now to the proof of

<u>Theorem</u> 2. For a numerical Campedelli surface S $\Phi = \Phi_{3K}$ is a birational map.

<u>Proof</u>. Consider $\Phi : S$ —— $\mathbb{P}^6$: $V = \Phi(S)$ is not contained in any hyperplane so that $d = \deg V \geq 5$.
V is not a curve, otherwise the general element of $|3K|$ would be decomposable in more than d elements, while we know that the curves D with $K\cdot D \leq 1$ are a finite number.

By Theorem 5.1 of [2] (also [7] Th. A) we know that $|3K|$ has no base points, hence if $m=\deg \Phi$, $dm = (3K)^2 = 18$.

We must then prove that it is impossible to have either $\deg \Phi = 2$ or $\deg \Phi = 3$.

<u>Case</u> I. $\deg \Phi = 2$

There is defined on S a birational involution σ such that $y=\sigma(x)$ if $\Phi(x) = \Phi(y)$; S being a minimal model σ is an automorphism, hence $\sigma^*(\mathcal{O}(K)) = \mathcal{O}(K)$.

Remark that $\sigma(C) = C$: in fact if a is a general point of C, and b is the point of C s.t $\mathcal{O}_C(a+b) \cong h_C$, $\Phi(a) = \Phi(b)$ (by lemma 5).

The exact sequence $0 \longrightarrow H^{\circ}(\mathcal{O}(-K)) \longrightarrow H^{\circ}(\mathcal{O}(K)) \longrightarrow$ $\longrightarrow H^{\circ}(\mathcal{O}_C(K)) \longrightarrow 0$ implies $h^{\circ}(\mathcal{O}_C(K)) = 0$, which is impossible by virtue of the following.

Lemma 6. $\mathcal{O}_C(K) \cong h_C^{\otimes 2}$.

Proof of the lemma. By lemma 5 $\mathcal{O}_C(3K) \cong h_C^{\otimes 6}$ so it suffices to prove that, for instance, $\mathcal{O}_C(8K) \cong h_C^{\otimes 16}$. But $|4K|$ has no base points, and we may pick up $s \in H^{\circ}(\mathcal{O}_S(4K))$ with 16 simple zeros on C, $a_1, \ldots a_{16}$. Then $\sigma^* s$ has 16 simple zeros too, $\sigma(a_1), \ldots \sigma(a_{16})$, therefore $\sigma^*(s) \cdot s|_C$ is a section of $\mathcal{O}_C(8K)$ whose divisor is linearly equivalent to $h_C^{\otimes 16}$.

Q.E.D.

Case II: $\deg \Phi = 3$.

Let a be a general point of C, and $\bar{a}$ be such that $a + \bar{a} \in |h_C| : \Phi(a) = \Phi(\bar{a})$, and the third point a' with the same image under Φ cannot lie on C. Set $N = \Phi(C)$: then $\Phi^{-1}(C) = C \cup N'$ and N' is a rational curve (there is a birational map from $N' = \text{locus}\{a'\}$ and $|h_C| \cong \mathbb{P}^1$). So S contains a continuous family of rational curves, which is though absurd.

Q.E.D.

V. Birationality of Φ_{3K} for a numerical Godeaux surface.

Denote by $\Phi = \Phi_{3K_S}$ and by $V = \Phi(S)$, $d = \deg V$.

Lemma 7. V is not a curve.

Proof. Otherwise d would be ≥ 3 and the moving part of $|3K|$ would be decomposable in more than d elements, which, by lemma 1, have intersection with K at least 2: then one would have $3K\cdot K \geq 2d \geq 6$.

Lemma 8. $|3K|$ has no fixed part.
See [6] , pag. 103 for the proof.

However $|3K|$ can have base points, and they are characterized by the following proposition, in which we denote, as in the following, by $T = \text{Tors}(\text{Pic}(S))$ and by D_ε, if $\varepsilon \neq 0$, $\varepsilon \in T$, the unique curve in $|K+\varepsilon|$.

Proposition 9. If b is a base point of $|3K|$ there exists $\varepsilon \in T$, $\varepsilon \neq -\varepsilon$, such that $C=D_\varepsilon + D_{-\varepsilon}$ is the unique curve in $|2K|$ passing through b : moreover D_ε and $D_{-\varepsilon}$ have b as the unique point of transversal intersection. Conversely, if $\varepsilon \neq -\varepsilon$, $D_\varepsilon \cap D_{-\varepsilon}$ gives a base point of $|3K|$.

Proof. We recall first Miles Reid's lemma (see Dolgacev's lecture) which asserts that if $\varepsilon \neq \tau$ are non zero torsion classes, D_ε and D_τ have no common component, hence intersect transversally in only one point.

Take $C \in |2K|$ such that $b \in C$: then $\mathcal{O}_C(3K) \cong \omega_C$ and b is a base point of $|\omega_C|$ (because $h^1(\mathcal{O}(K))=0$, so $|3K|\big|_C = |\omega_C|$). $2p(C)-2 = (C+K)\cdot C=6 \Rightarrow p(C)=4$ and by lemma B C must be reducible; moreover if Γ is a component of C containing b, $K\cdot \Gamma \geq 1$ ($K\cdot\Gamma=0 \Rightarrow \Gamma$ is rational non singular, hence that every section of $\mathcal{O}(3K)$, as it vanishes at b, vanishes on the whole of Γ : this would contradict lemma 8), so that b belongs to at most two components. Then pick up an irreducible component

$\Gamma_\circ$ of C such that, if $C=\Gamma_\circ+C_\circ$, b belongs to one and only one component of $C_\circ$. Consider the exact sequence:

$$0 \longrightarrow \mathcal{O}(K+\Gamma_\circ) \longrightarrow \mathcal{O}(3K) \to \mathcal{O}_{C_\circ}(3K) \longrightarrow 0 \ .$$

By Serre's duality $h^i(\mathcal{O}(K+\Gamma_\circ))=h^{2-i}(\mathcal{O}(-\Gamma_\circ))=0$ for $i=0,1$ ($i=1$ is a consequence of the exact sequence $0 \to \mathcal{O}(-\Gamma_\circ) \to \mathcal{O} \longrightarrow \mathcal{O}_{\Gamma_\circ} \longrightarrow 0$ and the irreducibility of $\Gamma_\circ$) and one obtains that b is a base point for $|\mathcal{O}_{C_\circ}(3K)|$.

Then one has the exact sequence

$$0 \to \mathbb{K} \to H^1(\mathfrak{m}_b\mathcal{O}_{C_\circ}(3K)) \text{ --- } H^1(\mathcal{O}_{C^\circ}(3K)) \text{ --- } 0$$

and, dualizing,

$$(*) \quad 0 \leftarrow \mathbb{K} \leftarrow \mathrm{Hom}(\mathfrak{m}_b, \mathcal{O}_{C^\circ}(-\Gamma_\circ)) \leftarrow H^\circ(\mathcal{O}_{C^\circ}(-\Gamma_\circ)) \longleftarrow 0$$

First, b must be a simple point of $C_\circ$, otherwise lemma B' implies $h^\circ(\mathcal{O}_{\tilde{C}_\circ}(-\Gamma_\circ)) \geq 1$, and using Lemma A plus the already quoted connectedness theorem for pluricanonical divisor, $\mathcal{O}_{\tilde{C}_\circ}(-\Gamma_\circ)$ has degree ≤ -1, hence there exists a decomposition of $C_\circ = C_1+C_2$ such that $-\Gamma_\circ \cdot C_2 \geq C_1C_2$: however then $0 \geq (\Gamma_\circ+C_1)\cdot C_2$, and $C=(\Gamma_\circ+C_1)+C_2$ is not numerically connected. The same reasoning gives the vanishing of $h^\circ(\mathcal{O}_{C_\circ}(-\Gamma_\circ))$, and b being simple, $(*)$ amounts to $h^\circ(\mathcal{O}_{C_\circ}(-\Gamma_\circ+b))=1$. Again lemma A gives either $\mathcal{O}_{C_\circ}(\Gamma_\circ)\cong\mathcal{O}_{C_\circ}(b)$ or $C_\circ=C_1+C_2$ such that $-\Gamma_\circ C_2+1 \geq C_1C_2$ if $b \in C_2$

$$-\Gamma_\circ C_2 \geq C_1C_2 \quad \text{if} \quad b \notin C_2 \ .$$

This last is impossible, the other two possibilities imply $C=C'+C''$ where $C'\sim K\sim C''$, again by the connectedness theorem , and $\mathcal{O}_{C'}(C'')=\mathcal{O}_{C'}(b)$ by lemma A again.

Finally the exact sequence

$$0 \longrightarrow H^{\circ}(\mathcal{O}(C''-C')) \longrightarrow H^{\circ}(\mathcal{O}(C'')) \longrightarrow H^{\circ}(\mathcal{O}_{C'}(C'')) \longrightarrow 0$$

implies that $C' \neq C''$, so $C'=D_{\varepsilon}$ for a suitable $\varepsilon \neq -\varepsilon$. Conversely, if $b=D_{\varepsilon} \cap D_{-\varepsilon}$ we claim that b is a base point for $|3K|$. In fact $h^1(\mathcal{O}(3K-D_{\varepsilon})) = h^1(-D_{-\varepsilon})=0$ (the $D_{\tau}'s$ are $\sim K$ hence numerically connected), so that $|3K||_{D_{\varepsilon}} =$ $= |\mathcal{O}_{D_{\varepsilon}}(3K)|$: then, as $\mathcal{O}_{D_{\varepsilon}}(3K) \cong \omega_{D_{\varepsilon}} \otimes \mathcal{O}_{D_{\varepsilon}}(b)$, $h^1(\mathcal{O}_{D_{\varepsilon}}(3K)) =$ $= h^{\circ}(\mathcal{O}_{D_{\varepsilon}}(-b)) = 0$, so b is a base point of $|3K||_{D_{\varepsilon}}$, and therefore of $|3K|$.

Q.E.D.

Lemma 10. If there exists $\varepsilon \in T$ s.t. $\varepsilon \neq -\varepsilon$, then $|3K|$ is spanned by the two linear subsystems $D_{\varepsilon} + |2K-\varepsilon|$, $D_{-\varepsilon} + |2K+\varepsilon|$.

Proof. Note that by R.R. $\forall$ non zero torsion class τ $h^{\circ}(\mathcal{O}(2K+\tau))=2$, while $P_3=4$, hence it suffices to show that these two subsystems have no common element. This is clear, however, since if one should have $D_{\varepsilon} + M_{-\varepsilon} = D_{-\varepsilon} + M_{\varepsilon}$, $M_{\varepsilon} - D_{\varepsilon}$ would be a positive divisor $\equiv K$ (in fact D_{ε} and $D_{-\varepsilon}$ have no common component), contradicting $p_g=0$.

Q.E.D.

Proposition 11. Two general curves of $|3K|$ are simple at a base point b of $|3K|$, and have there a transversal intersection.

Proof. If a general curve of $|3K|$ would be singular at $b=D_{\varepsilon} \cap D_{-\varepsilon}$, b would then be a double base point of $|3K||_{D} =$ $= b+|\omega_{D_{\varepsilon}}|$, so b would be a base point of the canonical system of D_{ε} ; so if Γ is the component of D to which b belongs $p(\Gamma)=0$ and, by Lemma A, $D_{\varepsilon} = C_1+C_2$, where $C_1 \geq \Gamma$, $C_1 \cdot C_2 \leq$ $\leq C_2(K+D_{\varepsilon})=2K\cdot C_2=0$ (as C_2 is made out of curves E s.t. $K\cdot E=0$): this however contradicts the numerical connectedness of D_{ε}. This reasoning tells also that a general curve is not tangent at b neither to D_{ε}, nor to $D_{-\varepsilon}$. If b is not a base point of $|2K-\varepsilon|$, or is not a base point of $|2K+\varepsilon|$, we are through: but in the contrary case, by lemma 10, b would be a singular base point of $|3K|$, which we have just shown to be impossible.

Q.E.D.

Denote by n the order of the torsion group T of S : by theorem 14 of [C.M.] (pag. 214-5) $n \leq 6$; moreover $n=6 \Longrightarrow$

$\mathrm{Tors}(\mathrm{Pic}(S)) \cong \mathbb{Z}/_{2\mathbb{Z}} + \mathbb{Z}/_{3\mathbb{Z}}$, hence there would exist a double unramified cover $p:\tilde{S} \to S$, with $\chi(\mathcal{O}_{\tilde{S}})=2$, $K^2_{\tilde{S}}=2$, $q(\tilde{S})=0$ ([C.M.] lemma 14, pag. 212), but then $\tilde{T}=\mathrm{Tors}(\mathrm{Pic}(\tilde{S}))$ should be either 0 or $\mathbb{Z}/_{2\mathbb{Z}}$ ([C.M.] th. 15, pag. 215).

Hence $n \leq 5$, and $\mathbb{Z}/_{2\mathbb{Z}} + \mathbb{Z}/_{2\mathbb{Z}}$ cannot be the torsion group, by Miles Reid's lemma (compare Dolgacev's lecture).

Combining these with the previous results, we obtain.

<u>Corollary</u> 12. There are no infinitely near base points for $|3K|$, and the number b of them is

$$b = \begin{cases} 0 & \text{if } T \cong 0, \mathbb{Z}/_{2\mathbb{Z}} \\ 1 & \text{if } T \cong \mathbb{Z}/_{3\mathbb{Z}}, \mathbb{Z}/_{4\mathbb{Z}} \\ 2 & \text{if } T \cong \mathbb{Z}/_{5\mathbb{Z}} . \end{cases}$$

Theorem 3. Φ is birational.

We refer the reader to [6], pag. 107-108 for the proof of this last part. We only remark that one has to prove that $m = \deg \Phi$ cannot be more than 1, and so one must show that (as $9 = md + b$, and $d \geq 2$) the following cases cannot hold

i) $d = 2 \quad m = 4 \quad b = 1$

ii) $d = 3 \quad m = 3 \quad b = 0$

iii) $d = 4 \quad m = 2 \quad b = 1$

For case i) it suffices to consider that V, a quadric, contains a pencil of reducible hyperplane sections, and by taking inverse images one contradicts lemma 1.

Case ii) is managed showing that V cannot have a double line (by a similar argument to the preceding one), hence it is a normal cubic, and then that there exists a pencil of quadrics cutting on V the images of curves in $|2K|$: however this gives rise to a numerical contradiction.

Finally case iii) makes direct use of the existence of the divisors D_{ε} homologous to K (guaranteed by corollary 12).

REFERENCES

[1] ≅ [C.M.] Bombieri, E. Canonical models of Surfaces of General Type, Publ. Math. I.H.E.S. 42 , pp. 171-219.

[2] Bombieri, E. The Pluricanonical Map of a Complex Surface, Several complex Variables I, Maryland 1970 Springer Lecture Notes 155, pp. 35-87.

[3] Bombieri, E.-Catanese, F. The tricanonical map of a surface with $K^2=2$, $p_g=0$, Preprint Pisa (to appear in a volume dedicated to C.P. Ramanujam).

[4] Bombieri, E-Catanese, F. Birationality of the quadricanonical map for a numerical Godeaux surface (to appear in B.U.M.I).

[5] Kodaira, K. Pluricanonical systems on algebraic surfaces of general type, J. Math. Soc. Japan, 20 (1968) pp. 170-192.

[6] Miyaoka, Y. Tricanonical maps of numerical Godeaux Surfaces Inventiones Math. 34 (1976), pp. 99-111.

[7] Miyaoka, Y. On numerical Campedelli Surfaces, Complex Analysis and Algebraic Geometry, Cambridge Univ. Press 1977, pp. 113-118.

[8] Mumford, D. The canonical ring of an algebraic surface, Annals of Math., 76(1962), pp. 612-615.

[9] Mumford, D. Lectures on Curves on an algebraic surface, Annals of Math. Studies, 59 (1966).

[10] Peters, C.A.M. On two types of surfaces of general type with vanishing geometric genus, Inventiones Math. 32, (1976). pp. 33-47.

[11] Ramanujam, C.P. Remarks on the Kodaira vanishing theorem, J. Indian Math. Soc. 36 (1972), pp.41-51.

CENTRO INTERNAZIONALE MATEMATICO ESTIVO

(C.I.M.E.)

ON A CLASS OF SURFACES OF GENERAL TYPE

F. CATANESE

ON A CLASS OF SURFACES OF GENERAL TYPE

F. Catanese(*)

Pisa

I. Introduction.

This lecture contains an exposition, without many details and proofs, (they will appear in a future paper), of a joint research of E.Bombieri-F.Catanese, dealing whith surfaces having the following numerical invariants: $K^2=2$, $p_g=q=1$ (of course they are of general type).

The interest about the existence of these surfaces was motivated by thefollowing remark: if S is a minimal surface of general type, and one considers the numerical characters K^2, $\chi = \chi(\mathcal{O}_S) = 1-q+p_g$, then the following inequalities hold

(*) Research made when this author was a member of G.N.S.A.G.A. of C.N.R..

(Noether's inequality) $p_g \le \frac{1}{2} K^2 + 2 \Rightarrow \chi \le \frac{1}{2} K^2 + 3$

(Bogomolov-Miyaoka's inequality) $K^2 \le 9 \chi$.

(a consequence of Castelnuovo's criterion) $\chi \ge 1$.

Moreover, if S is irregular, S admits unramified covers $\hat{S}$ of any order m ; as these two invariants for $\hat{S}$ are those of S multiplied by m , and Noether's inequality must hold for then, it turns out that $K_S^2 \ge 2 \chi (\mathcal{O}_S)$. In the (χ, K^2) plane then, the minimal irregular surfaces of general type lie in a convex ragion whose lower vertex corresponds to surfaces with $K^2=2$, $\chi=1$.

Finally in this case one must have $q=p_g=1$, by the results of [C.M.] (pag. 212), and one is conducted to check if these minimal values of K^2, χ are really attained, trying to construct such surfaces.

We have proved firstly an existence and unicity theorem about these surfaces, namely that there exist double ramified covers of the double symmetric product of an elliptic curve which have these numerical invariants, and moreover that all such surfaces arise in this way.

It has been then possible to show that their canonical models all belong to a family with non singular base space, hence by the results of [11] they are all deformation of each other (with non singular base), and are all diffeomorphic; their fundamental group is proven to be abelian exploiting this remark and a suitable degeneration of the branch locus: finally it is possible to prove that the constructed family coincides with the Kuranishi family (when the canonical models are nonsingular).

II. Geometry of the double symmetric product of an elliptic curve.

Let E be an elliptic curve; $E^2 = E \times E$: $E^{(2)}$, the double symmetric product of E is the quotient of E^2 by the involution taking (x,y) E^2 to (y,x).

The quotient map $\pi : E^2 \longrightarrow E^{(2)}$ has the diagonal Δ of E^2 as ramification locus, and $E^{(2)}$ is non singular (compare [1]). Fixing a base point in E as the zero element of a group law on E, the map $p' : E^2 \longrightarrow E$ s.t. $p'((x,y)) = x+y$ induces a mapping $p : E^{(2)} \longrightarrow E$ (note that a different choice of the base point alters p only up to a translation on E).

The following hold:

Proposition 1. $p : E^{(2)} \longrightarrow E$ makes $E^{(2)}$ into a $\mathbb{P}^1$-bundle over E.

Proposition 1 bis. p possesses a 1-dimensional family of cross sections, two of which intersect transversally in only one point.

Proof. For $a \in E$ consider the section $\sigma_a : E \longrightarrow E^{(2)}$ s.t. $\sigma_a(z) = (a, z-a)$. Denoting by Γ_a the image of σ_a, Γ_a and $\Gamma_{a'}$ meet transversally in (a,a').

Q.E.D.

Denoting by γ the homology class of a section and by f that of a fibre, we know that the second homology group of $E^{(2)}$ is freely generated by γ, f and $\gamma \cdot f = 1$, $\gamma^2 = 1$, $f^2 = 0$ (the first assertion is by the Leray-Hirsch theorem).

If D is a divisor, denote by $\deg D$ the intersection num ber of D with a fibre; one has

Proposition 3. The following sequence is exact

$$0 \longrightarrow \mathrm{Pic}(E) \xrightarrow{P_*} \mathrm{Pic}(E^{(2)}) \xrightarrow{\deg} \mathbb{Z} \longrightarrow 0 .$$

The proof of the preceding assertion is by diagram chasing, while, by the fact that every morphism of P^n into an abelian variety is constant, one gets.

Proposition 4. If D is a positive divisor on $E^{(2)}$, $D^2 \geq 0$.

We come now to the canonical bundle of $E^{(2)}$: considering the map π and the triviality of K_{E^2}, we obtain the equality $-2K = \hat{\Delta}$, where $\hat{\Delta} = \pi_* \Delta$.

For $z \in E$ denote by $F_z = p^{-1}(z)$; using prop. 3 and the structure of Pic (E) we can show that for a suitable choice of the base point in E $K = -2\Gamma + F_o$.

As for positive divisor D, $\deg (D) \geq 0$, one can sharpen prop. 4 to

Proposition 5. If B is an irreducible curve with $B^2=0$, either i) B is a fibre

or ii) B is homologous to $-nK$, $n \geq 1$.

There are on $E^{(2)}$ two rational pencils of curves, which will be of particular use in the sequel, that we are just going to describe.

Consider on E^2 the following family of curves, $\{C_a\}_{a \in E}$, given by $C_a = \{(x,x+a)\ x \in E\}$: these curves are isomorphic to E and are "translated" of $\Delta = C_o$.

Observe that $C_a \cap C_b = \emptyset$ if $a \neq b$, and the symmetry involution takes C_a to C_{-a} (which is a different curve if

$2a \neq 0$). Denote by a_i, $i=1,2,3$, one of the three points such that $a \neq 0$ $2a=0$: the symmetry operation operates trivially on Δ, not on $C_{\hat{a}_i}$.

If $\hat{C}_i = \pi(C_{\hat{a}_i})$ $\pi: C_{\hat{a}_i} \longrightarrow \hat{C}_i$ is a double unramified cover, so $\hat{C}_i$ is an elliptic curve not isomorphic to E, $\pi_*(C_{\hat{a}_i}) = 2\hat{C}_i$ while for the other values of a

$$\pi : C_a \longrightarrow p(C_a) \text{ is an isomorphism.}$$

We can summarize this in the following

Proposition 6. There exists on $E^{(2)}$ a rational pencil of curves linearly equivalent to $-2K$, whose elements are all isomorphic to E, but three particular members which are twice a non singular elliptic curve not isomorphic to E.

Proof. The family is given by $\{\pi_*(C_a)\}_{a \in E}$.
As $\pi_*(C_a) = \pi_*(C_{-a})$ the family can be parametrized by $\mathbb{P}^1$: then recall that $-2K \equiv \hat{\Delta} \equiv \pi_*(C_o)$.

Q.E.D.

By an entirely equal argument we have

Proposition 7. $|2\Gamma|$ contains the rational pencil given by

$$\{\Gamma_a + \Gamma_{-a}\}_{a \in E}.$$

Proposition 8. On $E^{(2)}$ the three curves $\hat{C}_i$ are the only curves homologous to $-K$.

Sketch of proof. If C is such a curve, $\hat{\Delta} \cdot C=0$, so

$\hat{\Delta} \cap C = \emptyset$: by an easy numerical argument one gets that $\pi^{-1}(C)$ is a connected curve $\tilde{C}$. However, $\tilde{C}$ being homologous to Δ and disjoint from it, it must be the graph of a translation on E (cf. [10]), so one of the curves C_a: finally the condition $\pi^* \pi_*(\tilde{C}) = 2\tilde{C}$ implies $\tilde{C} = C_{\hat{a}_i}$ for some $i \in \{1,2,3\}$.

Corollary 9. $|-3K| \neq \emptyset$, $h^0(-nK) = h^1(-nK)$, for $n \geq 0$.

Proof. Consider that $\hat{C}_i \equiv 2\Gamma - F_{\hat{a}_i}$, so that the sum of those three curves is $\equiv -3K$.
The other assertion comes from R.R. plus duality.

Q.E.D.

Proposition 10.

$$h^0(\mathcal{O}(-rK)) = \begin{cases} m+1 & \text{if } r=2m \\ m & \text{if } r=2m+1 \end{cases}$$

Proof. The proof of the first assertion uses induction on m in the cohomology sequence of

$$0 \longrightarrow \mathcal{O}(m(-2K)) \longrightarrow \mathcal{O}((m+1)(-2K)) \longrightarrow \mathcal{O}_{\hat{\Delta}} \longrightarrow 0 ,$$

to obtain $h^0(\mathcal{O}(-2mK)) \leq m+1$, and exploits the fact that $|-2mK| \supset m|-2K|$ to deduce that $h^0(\mathcal{O}(-2mK)) \geq m+1$.

For the second one we apply Ramanujam's theorem A ([9] [C.M.]): $|-2mK|$ being composed of the rational pencil $|-2K|$, $h^1(\mathcal{O}(2mK)) = m-1$.

Then Serre duality and corollary 9 are enough to fulfill the proof.

Q.E.D.

If τ is an automorphism of E, $(x,y) \longrightarrow (\tau(x),\tau(y))$ defines an automorphis $K(\tau)$ of $E^{(2)}$; any automorphism g of

$E^{(2)}$ moreover has the effect of permuting the fibres of p and so induces an automorphism $H(g)$ of E. A simple calculation shows that $H(K(\tau))(x) = \tau(x)+\tau(0)$ so that Ker $H\circ K$ is given by the four translations of period two. Then we observe that, as $(a,b) = \Gamma_a \cap \Gamma_b$ and $|\Gamma_a| = \{\Gamma_a\}$, if g^* is the identity on $\mathrm{Pic}(E^{(2)})$, g must be the identity ; so if $g \in \mathrm{Ker}\, H$, $g^*(F_a)=F_a$, $g^*(K)=K$, and $g^*(2\Gamma)=g^*(-K+F_o)=2\Gamma$.

Hence $g^*(\Gamma) = \Gamma - F_o + F_{\hat{a}_i}$, and Ker H has order 4.

Then, noticing that K is injective, Ker $H\circ K$=Ker H, and $H\circ K$ is onto, one gets.

Proposition 11. H is an isomorphism of $\mathrm{Aut}(E)$ onto $\mathrm{Aut}(E^{(2)})$.

III. Existence of surfaces with $K^2 = 2$, $p_g=q=1$.

Given a line bundle L on $E^{(2)}$, suppose we are given with a cover $\{U_i\} = \mathcal{U}$, a cocycle $(r_{ij}) \in H^1(\mathcal{U}, \mathcal{O}^*)$ defining L, and a positive divisor D defined by a section $S=\{s_i\}$ of $L^{\otimes 2}$: then one may take in L the surface S' defined by taking the square roots of S, that is S' is defined in $U_i \times \mathbf{C}$ (with coordinates (x,z_i)) by the equation $s_i(x)=z_i^2$. If $D=\mathrm{div}(S)$ is smooth, then S' is smooth too, and S' is normal if D has no multiple components.

Supposing that D exists, smooth, one obtains, using the signature formula of 4 and other computations, that, L being $\equiv$ $\equiv a\Gamma + ((b-1)F_o + F_x)$, S' has the desired numerical invariants iff $a=3$, $b=1$.

So we are left with the task of showing that the generic element of $|D|$ is non singular if $D \equiv 6\Gamma - 2F_x \equiv -3K+F_{-2x}$; in view of the result of prop. 11 one can restrict himself to consider only the divisor $D \equiv -3K+F_o$.

Proposition 12. $h^1(\mathcal{O}(D))=0$, $|D|$ has dimension 6 and its generic element is irreducible non singular.

Sketch of proof. as $|D| \supset |-2K| + \hat{C}_i + F_{-\hat{a}_i}$ $(i=1,2,3)$, $|D| \ni \hat{C}_1+\hat{C}_2+\hat{C}_3+F_o$ it is easily seen that $|D|$ has neither fixed part nor base points, and Bertini's theorem applies. The first two assertions follow from R.R., Serre duality and Ramanujam's theorem $(h^1(\mathcal{O}(D)) = h^1(-4K+F_o))$.

By the results of [5] , when D has no multiple components, S' has at most rational double points as its singularities iff D has no singular points of multiplicity greater than three, and any triple point has no infinitely near singularities of multiplicity greater than two; moreover if this conditions are satisfied , a minimal desingularization $g:S \longrightarrow S'$ of S' has the desired numerical invariants.

We observe that the condition is fulfilled in our linear system if D is irreducible (by the genus formula), and it is possible to describe explicitly which are the reducible curves in $|-3K+F_o|$, and which the "bad" ones: we omit this point for the sake of brevity, as well as the verification of

Proposition 13:

i) S is a minimal surface

ii) S' is the canonical model of S (see [C.M.] for its definition and properties)

iii) if $f: S' \longrightarrow E^{(2)}$ is the projection induced from L,
$h = p \circ f \circ g : S \longrightarrow E$ is the Albanese mapping of S.

We have moreover

Proposition 14. Let S, S_0 be two surfaces obtained in the above described way: if $\Psi: S \longrightarrow S_0$ is an isomorphism, $E \cong Alb(S) \cong Alb(S_0) \cong E_0$ and under this identification of E and E_0 Ψ is induced by an automorphism ψ of $E^{(2)}$ taking L to L_0, D to D_0.

Sketch of proof. By proposition 13 you may identify E, E_0 and suppose that Ψ commutes with the respective Albanese maps of S, S_0.
Ψ induces an isomorphism ψ' of the canonical models S', moreover the fibres $\hat{F}$ of the Albanese map are curves of genus two, and $f(x)=f(y)$ iff $\mathcal{O}_{\hat{F}}(x+y)$ is the unique hyperelliptic bundle of $\hat{F}$: this remark enables us to define ψ on $E^{(2)}$.

IV. Surfaces with $K^2=2$, $\chi=1$, $q=1$ are double covers of the symmetric product of their Albanese variety.

Let $\vartheta: S \longrightarrow E = Alb(S)$ be the Albanese map of S, and, for $u \in E$, set $G_u = \vartheta^{-1}(\{u\})$, $K_u \equiv K + F_u - F_0$.
We refer to [C.M.] for the proof of the following facts

i) $\forall\, u \in E$ $h^0(S, \mathcal{O}(K_u))=1$ (denote then by C_u the unique curve in $|K_u|$).

ii) C_u is generically irreducible $p(C_u)=3$

iii) $h^0(\mathcal{O}(K+K_v)) = 3$ $\forall\, v \in E$.

Fixing $v \in E$, $\forall u \in E$ we have the divisor $C_u + C_{v-u} \in$
$\in |K+K_v|$.

Proposition 15. The mapping $u \longrightarrow C_u + C_{v-u}$ defines an holomorphic mapping Ψ_v of E into $|K+K_v| \cong \mathbb{P}^2$.

Proposition 16. The image Δ_v of Ψ_v is an irreducible rational curve.

Proof. $C_u + C_{v-u} = C_{v-u} + C_{v-(v-u)}$, so Ψ_v is invariant by the involution which takes $u \longrightarrow v-u$, and whose quotient is isomorphic to $\mathbb{P}^1$ (denote now by $\varphi_v : \mathbb{P}^1 \longrightarrow |K+K_v|$ the induced mapping). Q.E.D.

Proposition 17. For general v Δ_v is non singular.

Sketch of proof. An uniformizing parameter on the universal cover of E induces a derivation D on each vector bundle, and, given a section σ, we use the classical notation σ' for $D(\sigma)$.
Take a cover $\{U_i\}$ of S on which $\mathcal{O}(K_u)$ is trivialized for u near $\hat{u}$: then, σ_u being the section of $\mathcal{O}(K_u)$ defining C_u, $\sigma_u = \{\sigma_{iu}\}$, where σ_{iu} depends holomorphically on u.
The condition that Ψ_v is not of maximal rank at $u=\hat{u}$ can be read as $\varphi_i = \dfrac{\sigma'_{i\hat{u}}}{\sigma_{i\hat{u}}} = \dfrac{\sigma'_{i(v-\hat{u})}}{\sigma_{i(v-\hat{u})}} + \lambda$, for λ a suitable constant independent of i.

Now, for general v, if $\hat{u} \neq v-\hat{u}$, this equality implies that φ_i is a regular function on U_i.

But then, if $f_{ij}(u)$ is a cocycle defining $\mathcal{O}(K_u)$, one gets after a simple computation that $\frac{f'_{ij}(\hat{u})}{f_{ij}(\hat{u})} = \varphi_i - \varphi_j$, hence it is a coboundary in $H^1(S, \mathcal{O})$. A similar conclusion can be drawn in general if φ_v is not of maximal rank at any point.

Denoting by $\lambda_{ij}(u)$ a cocycle for $\mathcal{O}_E(u-0)$ (relative to a covering $\{V_i\}$ compatible with $\{U_i\}$), we have that

$\frac{f'_{ij}(\hat{u})}{f_{ij}(\hat{u})} = \vartheta^* \left(\frac{\lambda'_{ij}(\hat{u})}{\lambda_{ij}(\hat{u})} \right)$: $\vartheta^* H^1(E, \mathcal{O}) \longrightarrow H^1(S, \mathcal{O})$ is however an isomorphism (compare [3]), so $\frac{\lambda'_{ij}(\hat{u})}{\lambda_{ij}(\hat{u})}$ is a coboundary. By the homogeneity of E under translation, it is possible to write for all u $\frac{\lambda'_{ij}(u)}{\lambda_{ij}(u)} = \varphi_i(u) - \varphi_j(u)$, with $\varphi_i(u) \in \Gamma(V_i, \mathcal{O})$.

Now, integrating on any path from 0 to u, one gets

$\frac{\lambda_{ij}(u)}{\lambda_{ij}(0)} = \frac{\exp(\int_0^u \varphi_i(t)dt)}{\exp(\int_0^u \varphi_j(t)dt}$, and so 0 and u should be two linearly equivalent points, which is clearly absurd.

Q.E.D.

The fact that $C_u \cdot C_v = 2$ and the irrationality of the pencil $\{C_u\}$ implies that this pencil has no base points hence

Proposition 18. $\forall v$ $|K+K_v|$ has no base points and Δ_v is an irreducible conic.

Proof. It suffices to show that $\deg \Delta_v = 2$ (Δ_v being irreducible), and yet only that $\deg \Delta_v \geq 2$ (Δ_v being generally non singular). Take then $C_u, C_{u'}$ general, meeting in two points

x,y : C_u+C_{v-u}, $C_{u'}+C_{v-u'}$ represent two points in the plane $|K+K_v|$ which lie in the intersection of Δ_v with the line $|K+K_v-x|$.

Q.E.D.

Proposition 19. For general $y \in S$, there exists a unique point $(u,u') \in E^{(2)}$ (with $u \neq u'$), s.t. $y \in C_u \cap C_{u'}$.

Proposition 20. The correspondence of prop. 19 extend to a holomorphic mapping $f': S \longrightarrow E^{(2)}$. Moreover f' factors as $S \xrightarrow{g} S' \xrightarrow{f} E^{(2)}$, where

a) S' is the canonical model of S, g the canonical mapping

b) f is a finite map of degree two

c) there exists a line bundle L on $E^{(2)}$ and a section $\sigma \in H^{\circ}(L^{\otimes 2})$ such that S' is isomorphic to the surface of the square roots of σ in L .

V. Some results on the structure of these surfaces and on their deformations.

Sharpening the result of prop. 20 one can show that it is possible to choose $L \equiv 3\Gamma - F_{\circ}$ (essentially by using $\mathrm{Aut}(E^{(2)})$).

We will sketch very rapidly the construction of a family containing all the canonical models S' of our surfaces: take $\mathcal{E} \longrightarrow H$ the local universal family of elliptic curves over the Siegel upper halfplane, and form its symmetric fibre product $\mathcal{E}^{(2)} \longrightarrow H$. All the invertible sheaves $\mathcal{O}_{E^{(2)}}(3\Gamma - F_{\circ})$ fit together to form an invertible sheaf $\mathcal{L}$ on $\mathcal{E}^{(2)}$ and one can choo se sections $s_o,\ldots s_6 \in \Gamma(\mathcal{E}^{(2)},\mathcal{L}^{\otimes 2})$ such that for all $t_o\ldots t_6$, $s_t = \sum t_i s_i$ defines a relative divisor $\mathcal{D}_t$ on $\mathcal{E}^{(2)} \longrightarrow H$ (we use the terminology of [8]). Then on $\mathcal{E}^{(2)} \times \mathbb{P}^6 \longrightarrow H\times \mathbb{P}^6$ is defined an invertible sheaf that we still denote by $\mathcal{L}$ and a relative divisor $\mathcal{D}$ linearly equivalent to $2\mathcal{L}$. Define $\mathcal{S} \hookrightarrow \mathcal{L}$ to be the variety defined in $\mathcal{L}$ by the square roots of a section defining $\mathcal{D}$: then we have a family $\mathcal{S} \longrightarrow H\times \mathbb{P}^6$ and an open dense subset V of $H\times \mathbb{P}^6$ such that the fibres over points of V are all the canonical models of the surfaces we are considering.

By using the result of Tyurina about local resolution of singularities of such families ([11]) we deduce that our surfaces are all diffeomorphic.

Proposition 21. $\pi_1(S) \xrightarrow{h_*} \pi_1(E)$.

Sketch of proof. By the above remark we may consider the particular surface for which $D=F_o+\hat{C}_1+\hat{C}_2+\hat{C}_3$.
Then we observe that by Van Kampen's theorem $\pi_1(S)\cong\pi_1(S')$; call $\hat{F}_o=f^{-1}(F_o)$, and observe that $S'-\hat{F}_o$ —— $E-\{0\}$ is a differentiable fibre bundle with fibre a smooth curve of genus two. Take U to be a small disk around 0 in E : $(p\circ f)^{-1}(U) = S'|_U$ is contractible to $\hat{F}_o$, which is simply connected, and we can apply again Van Kampen's theorem to the open sets $S'|_U$, $S'-\hat{F}_o$ in S', U and $E-\{0\}$ in E , together with the exact homotopy sequence of a bundle (where $u \in U-\{0\}$, $\hat{F}_u$ is the fibre of S' over u), to obtain the following diagram, commutative, exact in the columns and the rows , and which gives easily our result.

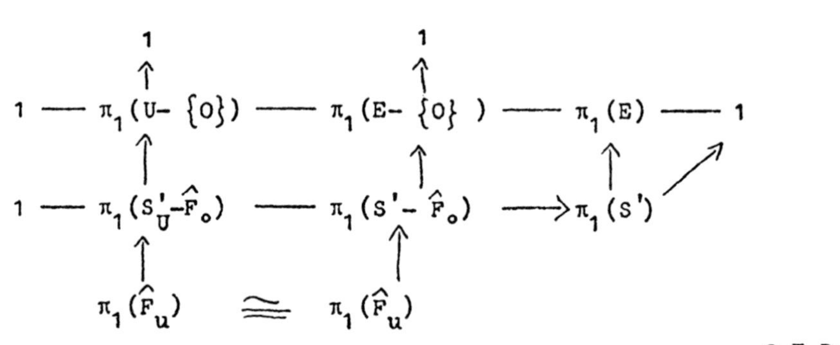

Q.E.D.

We end this talk by showing

Proposition 22. If θ_S denotes the tangent bundle of S, then $h^1(\theta_S)=7$.

Sketch of the proof (in the simpler case when D is smooth). Using the Hirzebruch Riemann - Roch theorem one sees that it is equivalent to show that $h^\circ(\Omega^1_S \otimes \Omega^2_S)=1$. One exploits first one fact: you have an involution on S determined by $f':S \longrightarrow X=E^{(2)}$, so $H^\circ(\Omega^1_S \otimes \Omega^2_S) = H^+ \times H^-$, where H^+ is the subspace of invariant, H^- of anti invariant sections of $\Omega^1_S \otimes \Omega^2_S$, and by one side the sections of H^+ correspond to sections of $\Omega^1_X \times \Omega^2_X \times L, \cong \Omega^1_X \otimes \mathcal{O}(\Gamma)$, by the other, sections of H^- vanish on the ramification locus R , hence they come from sections of $\Omega^1_S \otimes \mathcal{O}(K_S-R) \cong \Omega^1_S \times f^*(K_X)$. Secondly one exploits the unique section of Ω^1_X (which defines a trivial sub-bundle T of Ω^1_X) and the fact that $\Omega^1_X \otimes \mathcal{O}(\Gamma)$ $\Omega^1_X \cong \mathcal{O}(-\Gamma+F_\circ)$ has no non zero sections, to infer that every section of $\Omega^1_X \otimes \mathcal{O}(\Gamma)$ comes from a section of the subbundle $T \otimes \mathcal{O}(\Gamma) \cong$ $\cong \mathcal{O}(\Gamma)$, which, though, possesses only one section. The same method can be applied to show the vanishing of $H^{\circ}(\Omega^1_S \otimes f^*(K_X))$.

Q.E.D.

Finally, when D is non singular, one can take the constructed family and show the injectivity of the Kodaira-Spencer map: then locally our family is the Kuranishi family (compare [6], [7]).

References

[1] Andreotti, A. On a theorem of Torelli, Am. Jour. of Math. 80, pp. 801-828 (1958)

[2] [C.M.] Bombieri, E. Canonical models of surfaces of general type, Publ. I.H.E.S. 42, pp. 447-495 (1974)

[3] Chern, S.S. Complex manifolds, Universidad Recife (1959)

[4] Hirzebruch, F. The signature of ramified coverings, in "Global Analysis", Princeton Univ. Press 29, pp. 253-265 (1969)

[5] Horikawa, E. On deformations of quintic surfaces, Inv. Math. 31, pp. 43-85 (1975)

[6] Kodaira, K. - Morrow, J. Complex manifolds, Holt, Rinehart and Winston, Inc. (1971)

[7] Kuranishi, M. New proof for the existence of locally complete families of complex structures, in "Proceedings of the Conference on Complex Analysis in Minneapolis 1964", Springer-Verlag pp. 142-154 (1965)

[8] Mumford, D. Lectures on curves on an algebraic surface, Annals of Math. Studies, 59 (1966)

[9] Ramanujam, C.P. Remarks on the Kodaira vanishing theorem, Jour. Indian Math. Soc. 36, pp. 41-50 (1972)

[10] Siegel, C.L. Topics in complex function theory, Vol. II, Wiley - Interscience (1971)

[11] Tyurina, G.N. Resolution of singularities of plane deformations of double rational points, Funkcional Anal. i Prilozen 4, pp. 77-83 (1970)

CENTRO INTERNAZIONALE MATEMATICO ESTIVO

(C.I.M.E.)

SOME REMARKS ABOUT THE "NULLSTELLENSATZ"

A. TOGNOLI

SOME REMARKS ABOUT THE "NULLSTELLENSATZ".

A. Tognoli
(Pisa)

Introduction.

Let K be a field and Γ_{K^n} the ring of the regular functions (in the sense of F.A.C.) on K^n.

Two problems are now natural:

1) to carachterize the ideals of definition of Γ_{K^n}

2) to carachterize the ideals of definition of $K[X_1,\ldots,X_n]$.

We have the following results:

Theorem I. - An ideal I of Γ_{K^n} is of definition if and only if is intersection of maximal ideals.

<u>Theorem</u> II. - An ideal I of $K[X_1,\dots,X_n]$ is of definition if and only if for any homogenous polynomial P vanishing only in the origin we have :

$$P(h_1,\dots,h_q) \in I \iff h_1 \in I,\dots,h_q \in I \ .$$

In the following we shall prove theorem II (see [4]), theorem I is an easy consequence of proposition 1 .

<u>Preliminary remarks</u>

Let I be an ideal of $K[X_1,\dots,X_n]$, in what follows ideal shall mean "proper ideal" .

We shall call set of zeroes of I the set:

$$Z(I) = \{ x \in K^n \mid P(x) = 0 \ , \forall P \in I \} .$$

An ideal I is called <u>ideal</u> of <u>definition</u> if

$$I = \{ P \in K[X_1,\dots,X_n] \mid P|_{Z(I)} = 0 \} .$$

Let K be a field, we shall denote by $\mathcal{P}_K$ the set of the homogeneous elements P of $K[Y_1, Y_2]$ such that

$$P(Y_1,Y_2) = 0 \text{ if and only if } Y_1 = Y_2 = 0 \ .$$

Definition 1. - An ideal I of $K[X_1,\dots,X_n]$ is said to satisfy condition α if for any $P \in \mathcal{P}_K$ we have:

$$P(g_1,g_2) \in I \implies g_1 \in I, g_2 \in I .$$

It is a natural problem to see if $\mathcal{P}_K$ is empty. The following proposition gives the solution:

Proposition 1. - *Let* K *be a non algebraically closed field and* q N . *Then there exists a homogeneous polynomial* $P \in K[Y_1,\dots Y_q]$ *such that:*
0

1) $P(Y_1,\dots,Y_q) = 0$ *if and only if* $Y_1 = \dots = Y_q = 0$

2) *the coefficient of* Y_1^d *in* P *is* 1 , $d = \deg P$.

Proof. Let $\bar{K}$ be the algebraic closure of K and $K_s \supset K$, $K_s \subset \bar{K}$ the maximal separable extension of K .

It is well known (see [1]) that K_s exists and K is purely inseparable on K_s . Let $\omega \in K_s - K$ and suppose G the Galois group of a Galois extension of K containing ω .

If $P = Y_1 + \omega Y_2$ and $P_{12} = \prod_{g \in G} g(Y_1 + \omega Y_2)$ then P_{12} is invariant under G , and hence $P_{12} \in K[Y_1,Y_2]$.

Moreover we have: if $x \in K^2$ then $P_{12}(x) = 0 \iff Y_1 = Y_2 = 0$.

The polynomial P_{12} is homogeneous of degree, say d_1 , and the coefficient of $Y_1^{d_1}$ is 1.

In a similar way we define:

$$P_{123} = \prod_{g \in G} g(P_{12} + \omega Y_3^{d_1})$$

and after $q-1$ steps we obtain the desired result.

New suppose $K = K_s$, then there exists $\omega \in \bar{K} - K_s$. By hypothesis ω is purely inseparable on K and hence has a minimal polynomial of the form $X^{p^m} - \beta = 0$. In this case we define $P_{12} = (Y_1 + \omega Y_2)^{p^m}$ and by the above arguments we end the proof.

Corollary. If K is not algebraically closed $\mathcal{P}_K$ is not empty.

The proof of the theorem.

We wish to state the following

Theorem 1. Let K be a non algebraically closed field then an ideal I of $K[X_1, \ldots, X_n]$ is an ideal of definition if and only if it satisfies the condition α.

Proof. The part: I of definition $\Rightarrow I$ satisfies α is clear.

Now we shall prove the other implication.

Step. I. (reduction to prime ideals).

First we shall prove that if I satisfies α then $I = \sqrt{I}$.

Let $P \in K[Y_1, Y_2]$ be an element of $\mathcal{P}_K$ and let us suppose Y_1^d has 1 as coefficient, $d = \deg P$ (see proposition 1). We have: $P(Y_1, 0) = Y_1^d$ and this proves that if I satisfies α then $g^d \in I \Rightarrow g \in I$, and hence $I = \sqrt{I}$ (we remark that $g^q \in I \Rightarrow g^{nd} \in I$ if $n \gg 0$, but we can use $p^n \in \mathcal{P}_K$ ...).

Let now $I = \bigcap_j I_j$ be the decomposition of I as intersection of prime ideals.

We wish to prove that I satisfies α if and only if any I_j satisfies.

It is clear that if all the I_j satisfy α then I has the same property.

On the contrary suppose I satisfies α and I_1 does not satisfy.

Let $P(g_1, g_2) \in I_1$, $g_1 \notin I_1$.

If $g \in \bigcap_{j>1} I_j$, $g \in I_1$ we have:

$$g^d\, P(g_1, g_2) = P(gg_1, gg_2) \in I \;,\; d = \deg P \;. \tag{1}$$

By the fact that I satisfies α we have $gg_1 \in I_1$ and hence $g \in I_1$, or $g_1 \in I_1$ (I_1 is a prime ideal) and this is impossible.

So we have proved that it is enough to verify the theorem in the case I is a prime ideal that satisfies α .

Step 2. (The case $Z(I) = \emptyset$).

Let I be an ideal of $K[X_1,\dots,X_n]$ and suppose I satisfies α . We wish to prove that, if $Z(I) = \emptyset$, then $I = K[X_1,\dots,X_n]$. It is sufficient to prove that for any (proper) ideal I , maximal ámong those that satisfy α we have $Z(I) \neq \emptyset$.

By the above results we can suppose I prime.

By the proof of the normalization theorem (see [1]) we know that there exist $d_2,\dots,d_n \in N$ such that if:

$$(1) \qquad \begin{cases} Y_1 = X_1 \\ Y_2 = X_2 + X_1^{d_2} \\ \cdot \\ \cdot \\ \cdot \\ Y_n = X_n + X_1^{d_n} \end{cases}$$

then the class $\{Y_1\}$ in $K[X_1,\dots X_n]/_I$ is integral on $\{Y_2\},\dots,\{Y_n\}$.

Relations (1) are inverted by

$$
(2) \qquad \begin{cases} X_1 = Y_1 \\ X_2 = Y_2 - Y_1^{d_2} \\ \cdot \\ \cdot \\ \cdot \\ X_n = Y_n - Y_1^{d_n} \end{cases}
$$

Hence we have an isomorphism $\sigma_1 : K[X_1,\dots,X_n] \longrightarrow$
$\longrightarrow K[Y_1,\dots,Y_n]$.

Let trasc. $K[X_1,\dots,X_n]/_I = q$, then after $n - q$ steps we have an isomorphism $\sigma_{n-q} : K[X_1,\dots,X_n] \to K[W_1,\dots,W_n]$ where $\{W_1\},\dots,\{W_{n-q}\}$ are integral on $K[\{W_{n-q}\},\dots,\{W_n\}]$ and $\{W_{n-q+1}\},\dots,\{W_n\}$ is a base of trascendence of $\sigma_{n-q}(K[X_1,\dots,X_n])/\sigma_{n-q(I)}$.

Let us now identify $K[X_1,\dots,X_n]$ with $K[W_1,\dots,W_n]$ and I with $\sigma_{n-q}(I)$.

Let us denote $Z_1 = \{W_{n-q+1}\}\dots, Z_q = \{W_n\}$ and $\theta_1 = \{W_1\},\dots, \theta_{n-q} = \{W_{n-q}\}$.

We remark that I satisfies α if and only if, for any $a,b \in K[X_1,\dots,X_n]/I$ and $P \in \mathcal{P}_K$ we have: $P(a,b) = 0$ if, and only if, $a = b = 0$.

Let $\pi : K[X_1,\dots,X_n] \longrightarrow K[\theta_1,\dots, \theta_{n-q}, Z_1,\dots,Z_q]$ be the natural projection and $\mathfrak{m} = \pi^{-1}(\mathfrak{m}')$, where $\mathfrak{m}'$ is the ideal generated by $Z_1,\dots,Z_q$.

We wish to prove that $\mathfrak{m}'$ satisfies α and hence, by the maximality of I, we deduce that $q = 0$ and therefore $K[X_1,\dots,X_n]/_I$ is algebraic on K.

Before we prove that $\mathfrak{m}'$ satisfies condition α in $K[X_1,\dots,X_n]/_I$.

Let $P \in \mathcal{P}_K$, $P(g_1,g_2) \in \mathfrak{m}''$, $g_i = \sum_{j \geq 0} a^i_j Z^j$ (we use the multiindex j), $a^i_j \in K[\theta_1,\dots,\theta_{n-q}]$.

All the monomials of $P(g_1,g_2)$ that have positive degree are in $\mathfrak{m}'$, hence $P(g_1,g_2) \in \mathfrak{m}''$ if, and only if, $P(a^1_o, a^2_o) = 0$ where a^1_o and a^2_o are element of $K[\theta_1,\dots,\theta_{n-q}]$. I satisfies α hence we have: $P(a^1_o,a^2_o) = 0$ if and only if $a^1_o = a^2_o = 0$ and in these hypothesis $g_1 \in \mathfrak{m}''$, $g_2 \in \mathfrak{m}'$.

This proves that $\mathfrak{m}'$ satisfies the condition α in $K[X_1,\dots,X_n]/_I$.

Let now $h_1, h_2 \in K[X_1,\dots,X_n]$, $P \in \mathcal{P}_K$ and let us suppose $P(h_1,h_2) \in \mathfrak{m}$, then $P(\{h_1\}, \{h_2\}) \in \mathfrak{m}'$ where $\{h_i\}$ is the class of h_i modulo I.

But, by the above proof, we have: $h_i \in \mathfrak{m}'$ and hence $h_i \in \mathfrak{m}$ (because $\mathfrak{m} \supset I$).

We can now suppose $K[X_1,\dots,X_n]/_I$ is algebraic on K. In particular X_1 mod I is algebraic on K, hence there exists $P(X_1) = \sum_{i=o}^{q} a_i X_1^i \in I$, $a_i \in K$.

We may suppose $P(X_1)$ has no zero in K , because, on the contrary, we should be able to find an element of $Z(I)$ in K^n (see [1] pag. 197).

Let us consider the polynomial:

$$\hat{P}(X,t) = \sum_i a_i X^i t^{q-i} .$$

From the fact that P has no zero in K it follows that $\hat{P}(Y_1,Y_2) \in \mathcal{P}_K$.

We have $P(X_1) = P(X_1,1) \in I$ and, hence $1 \in I$ and this proves that we can not suppose $Z(I) = \emptyset$.

Step 3. (General case).

Let I be a prime ideal that satisfies α and $g \in K[X_1,\dots,X_n]$ such that $g|_{Z(I)} = 0$.

We must prove that $g \in I$.

Suppose $g \notin I$ and let us define $\varphi = 1-X_{n+1}g \in K[X_1,\dots X_{n+1}]$ Let us denote by $\tilde{I}$ the ideal of $K[X_1,\dots,X_{n+1}]$ generated by I and φ .

Let $\hat{I} = \{h \in K[X_1,\dots,X_{n+1}] \mid g^t h \in I$ for some $t \in N\}$.

It is not difficult to verify that $\hat{I}$ is an ideal and we wish to prove that $\hat{I}$ satisfies α .

Let $P \in \mathcal{P}_K$ and let us suppose $g_1, g_2 \in K[X_1, \ldots, X_{n+1}]$ such that:

$$(1) \quad g^s P(g_1, g_2) = \sum_i \alpha_i(X_1, \ldots, X_{n+1}) h_i(X_1, \ldots, X_n) + $$

$$+ \alpha(X_1, \ldots, X_{n+1}) \varphi$$

where h_i are some generators of I . Taking $\varphi = 0$, and hence $X_{n+1} = \frac{1}{g}$, from (1) we deduce:

$$(2) \quad P(g^r g_1, g^r g_2) = \sum_i \alpha_i^*(X_1, \ldots, X_n) h_i(X_1, \ldots, X_n)$$

where the α_i^* are obtained from the α_i by the substitution $X_{n+1} = \frac{1}{g}$ and by multiplication of a power of g to eliminate the denominators.

From (2) and the fact that I satisfies α we have $g_1 \in \hat{I}$, $g_2 \in \hat{I}$.

The ideal $\hat{I}$ has no zero ($\tilde{I}$ has no zero) and satisfies α , hence $1 \in I$.

We have therefore:

$$g^t = \sum_i \; {}_i(X_1, \ldots, X_{n+1}) h_i(X_1, \ldots, X_n) + \alpha(X_1, \ldots, X_{n+1}) \varphi$$

Now we restrict ourselves to $\varphi = 0$ and we obtain

$$g^{t+s} = \sum_i \alpha_i^*(X_1, \ldots, X_n) h_i(X_1, \ldots, X_n) \in I$$

and the theorem is proved because I is prime.

Remark 1. If we denote by $\mathcal{P}_K$ the set of homogeneous polynomials, in one or two variables, zero only in the origin, theorem 1 is true for any field K. In the following we shall assume this hypothesis.

Remark 2. Associating to any affine variety $V \subset K^n$ the cone in K^{n+1} we obtain the following:

Theorem. - A homogeneous ideal $I \subset K[X_1,\dots,X_{n+1}]$ is an ideal of definition if and only if I satisfies condition α.

Remark 3. Let $V \subset K^n$ be an affine variety and I_V the ideal of all polynomials zero on V.

We have:

Theorem. - An ideal $I \subset K[X_1,\dots,X_n]/I_V = P_V$ is an ideal of definition in P_V if, and only if, I satisfies condition α.

Proof. Let $\pi : K[X_1,\dots,X_n] \longrightarrow P_V$ be the natural projection, it is easy to see that $\pi^{-1}(I)$ satisfies α in $K[X_1,\dots,X_n]$ and the thesis follows.

Remark 4. Let I be a prime ideal of $K[X_i,\dots,X_n]$ that satisfies α then I is absolutely prime (id est the extension of I in any $\tilde{K}[X_1,\dots,X_n]$, K subfield of $\tilde{K}$, is prime).

Proof. $I \otimes_K \tilde{K}$ defines the completion of the variety V defined by I. If V is irreducible the completion is also irreducible (see [2]) and the result is proved.

Remark 5. Let $I \subset K[X_1,\dots,X_n]$ be an ideal that satisfies condition α then I is absolutely unramified (id est for any field $\tilde{K} \supset K$ we have $I \otimes_K \tilde{K} = \sqrt{I \otimes_K K)}$.

Proof. $I \otimes_K \tilde{K}$ defines the completion of the variety V defined by I (see [2]) and hence coincides with its radical.

Remark 6. Let K , $K \neq \bar{K}$, be a field, then a finitely generated K - algebra A , is isomorphic to the algebra of the polynomial functions restricted to an affine variety if, and only if,

$$P(g_1,g_2) = 0 \Longrightarrow g_1 = g_2 = 0, \quad g_1,g_2 \in A, \; P \in \mathcal{P}_K .$$

Remark 7. Let K be a real closed field and I an ideal of $K[X_1,\dots,X_n]$. Then I is an ideal of definition if, and only if:

β) $\quad g^2 + h^2 \in I \Longrightarrow g \in I,\ h \in I,\ g,h \in K[X_1,\dots,X_n]$.

Proof. Clearly if I is of definition β) is satisfied. We wish to prove that if β) is true then α is true. It is easy to verify that if I satisfies β then $I = \sqrt{I}$; using the arguments of theorem 1 we deduce that it is enough to prove the remark in the case I is a prime ideal.

Let $P \in K[Y_1,Y_2]$ be an homogeneous polynomial of $\mathcal{P}_K$, we can suppose P is irreducible of degree two (see [3]) .

If

$$P(Y_1,Y_2) = a\,Y_1^2 + b\,Y_1Y_2 + c\,Y_2^2 \ ,\ P \in \mathcal{P}_K$$

we have:

$$b^2 - 4ac < 0 \quad \text{and hence:}$$

$$(1) \qquad P(Y_1,Y_2) = (Y_1 + \frac{b}{2a}\,Y_2)^2 + (\frac{-b^2 + 4ac}{4a^2})Y_2^2 \ .$$

Let us suppose $P(g_1,g_2) \in I$ and β) is true, then from (1) it follows that $g_1 \in I$, $g_2 \in I$ and hence α is true and the remark is proved.

References

[1] L.BERETTA and A.TOGNOLI. "Some basic facts in algebraic geometry on a non algebraically closed field", Ann. Scuola Norm. Sup. Pisa, Ser. IV 3 (1976), 341-359.

[2] D.W.DUBOIS. "A nullstellensatz for ordered fields", Arkiv for Mat., 8 (1969), 111-114.

[3] J.J.RISLER. "Une caracterisation des ideaux des varietes algebriques reeles", C. R. Acad. Sci. Paris 271 (1970); 1171-1173.

[4] W.A.ADKINS, P.GIANNI, A.TOGNOLI. "A Nullstellensatz for an algebraically non-closed field". To appear on B.U.M.I..

Breinigsville, PA USA
26 November 2010
250071BV00003B/5/P